Population Ecology of Roosevelt Elk

Population Ecology of Roosevelt Elk

Conservation and Management in Redwood National and State Parks

BUTCH WECKERLY

UNIVERSITY OF NEVADA PRESS *Reno & Las Vegas*

University of Nevada Press | Reno, Nevada 89557 USA
www.unpress.nevada.edu

Cover design by Louise OFarrell

LIBRARY OF CONGRESS CATALOGING-IN-PUBLICATION DATA
Names: Weckerly, Butch, 1957- author.
Title: Population ecology of Roosevelt elk : conservation and management in Redwood National and State Parks / by Butch Weckerly.
Description: Reno : University of Nevada Press, [2017] | Includes bibliographical references and index.
Identifiers: LCCN 2017009452 (print) | LCCN 2017010424 (e-book) | ISBN 978-1-943859-50-4 (cloth : alk. paper) | ISBN 978-0-87417-782-4 (e-book)
Subjects: LCSH: Roosevelt elk—Ecology—Redwood National and State Parks (Calif.) | Roosevelt elk—Conservation—Redwood National and State Parks (Calif.) | Redwood National and State Parks (Calif.)—Conservation. | Redwood National and State Parks (Calif.)—Management.
Classification: LCC QL737.U55 W39 2017 (print) | LCC QL737.U55 (e-book) | DDC 599.65/70979412—dc23
LC record available at https://lccn.loc.gov/2017009452

The paper used in this book meets the requirements of American National Standard for Information Sciences—Permanence of Paper for Printed Library Materials, ANSI/NISO Z 39.48-1992 (R2002).

FIRST PRINTING

Manufactured in the United States of America

To

Chris, Mom and Dad

"The meaning of life is chasing elk.
The great mystery in life
is why everyone else does not think the same!"

Contents

Illustrations

FIGURES

TABLES

Preface

I was fitfully writing a new lecture in September 1994 when the technician for my elk study on private timberland, which was also new, called me. He was excited. In the process of conducting systematic surveys to locate recent elk sign (tracks and feces), he had decided to see if he could stalk an elk group. Over the phone he described how after he had encountered recent sign he patiently approached the elk, got a closer look, and, indeed, counted the elk in the group. What made this particularly challenging, and exciting, was that he was in a densely forested landscape where elk were difficult to spot. Furthermore, we could not afford expensive technology that would have made the task of locating elk easier. When this individual called, I was in a bad mood and so I was far from excited by his news. I quickly gave several reasons why stalking elk in thickly timbered areas was not going to be an effective way to count elk, said good-bye, and hung up the phone. In the end, what population survey protocol did we develop? One that relied on finding recent elk sign and stalking the group.

The most important outcome of developing the population survey protocol was my growing awareness that although I was the more experienced field biologist, I was also more dogmatic. If I was going to study elk in a heavily forested landscape with a small budget, I would have to be more imaginative, an approach that, by the way, benefits from a good attitude.

Shortly afterward I began studying elk in Redwood National and State Parks (RNSP). Again, I was working with a small budget. One of the early tasks was developing population survey protocols that are still used today. One survey protocol involves stalking and another protocol requires driving along the heavily used and scenic U.S. Highway 101. These are not very orthodox survey procedures that, nonetheless, appear to result in reliable estimates of abundance.

Developing efficient and effective techniques is rewarding. But, more importantly, why am I collecting data in the first place? In the RNSP elk are neither hunted nor scarce. In short, there are two primary reasons. One reason has to do with people. Elk are charismatic. Throughout the 20 years of this study I have been continually amazed at how elk are like magnets to people, continually drawing in people that look at them and then, quite often, photograph them at as close a distance as possible. This study provides information about the population ecology of Roosevelt elk that cannot be obtained in a brief visit to RNSP. The other reason has to do with understanding the ecological dynamics of a species that is a historical part of the fauna in the region. This understanding leads to knowledge that is necessary to manage elk in the RNSP so that they will be conserved for a long time, perhaps in perpetuity. Yet, how will elk persistence be assessed and what data are needed to manage persistence? In a nutshell, these are the problems and challenges confronted in my study that I discuss in this book.

Finally, I am indebted to the technician I employed more than 20 years ago. Why? He certainly showed me how to chase elk, not as easy a task as you might think. But more importantly, after I hung up the phone, he did not do what he could have done: find another employer that was less moody. To address wildlife persistence, what is first required is persistence by people.

Acknowledgments

Redwood National and State Parks (RNSP) personnel were integral to conducting my long-term study. My first RNSP contact was Howard Sakai. Through Howard I became familiar with other RNSP personnel involved in wildlife management. About six years into the study Howard retired and my new contact person became, and still is, Kristin Schmidt. Kristin is a wonderful person to work with; she provided access to National Park Service vehicles and set up my reservations with RNSP housing. In some years, reservations were easily set up and in other years there were last-minute issues that she attended to with professional calm. Sometimes she dealt with issues on Christmas or New Year's Eve.

There were other RNSP personnel that also provided help over the years: Leonel Arguello, Keith Benson, Heather Brown, Terry Hines, Kyle Max, Aida Parkinson, and David Roemer. Terry Hofstra was the supervisory ecologist in Redwood National Park for much of the time of the project. Terry had charisma and was always offering encouragement, particularly when I told him another grant proposal was rejected. From California State Parks, Jay Harris readily processed my annual permit applications. Jim Able gave me permission to survey Stover's Ranch. I had the pleasure of interacting with marvelous students on this project that did not shy away from hard work. Those students were Adam Duarte, Jason Hunt, Rene Keheler, Nicholas Kolbe, Meredith Longoria, Ryan Luna, Kim McFarland, Aaron McGuire, Michael O'Day, Leah Peterson, Mark Ricca, Katherine Richardson, Sarah Robinson, Stephanie Shelton, Heath Starns, Garrett Street, and Daniel Wolcott. Funding for the project was from Alamo Safari Club, California Fish and Wildlife, Granite Bay Safari Club, Rocky Mountain Elk Foundation, Houston Safari Club, and Save the Redwoods League.

Population Ecology of Roosevelt Elk

Introduction

To be able to describe something,
you have to feel some kind of emotional attachment to it.

— Karl Ove Knausgaard,
My Saga, Part I: On the Trail of the First Europeans to Set Foot in America

I did not know when and how it was going to happen but I did know that I was going to study these elk. This thought occurred in August 1993, while watching elk in Redwood National and State Parks (RNSP). At the time, I had been in the area for only 2 years. I was a full-time lecturer with a heavy teaching load and little time for what I truly wanted to do—research. My teaching load dropped a year later, which coincided with me initiating elk studies in various places in north-coastal California. One of the outcomes is this work, a case study of the dynamics of population distribution and abundance for the purpose of conserving and managing elk in a park setting. All of this was conducted on a shoestring budget.

At the time the project began, I could not afford a good pair of hiking boots. The shoestring label dawned on me when I realized one of the few things I could afford was replacement shoestrings for my high-top sneakers, my field boots. But in the wet, cool climate of north-coastal California—which had no rattlesnakes, by the way—a pair of high-top sneakers and wool socks functioned just as well as field boots. The sneakers provided ankle support and some grip on slippery logs. Wool socks kept my feet warm in the constant cool, soggy conditions. I was not wanting in the stuff needed to hike and do field work.

It is one thing to want to study something, but it is much more important to have reasons to study something (Lindenmayer & Likens, 2010). Without a need, the want recedes. At first glance there is not much need to provide information useful to managing elk in the RNSP. Elk are often labeled big game because they are hunted. Nevertheless, elk are not hunted in the RNSP and have not been since the creation of the RNSP. There is little need to estimate sustainable levels of harvest or to use an approach of learning by doing, an adaptive management framework, to estimate environmental influences on elk populations and harvests (McCullough, 1979; Nichols, Johnson, Williams, & Boomer, 2015). But hunting does occur in some areas adjacent to the RNSP, and the elk available to harvest probably do spend much time in the RNSP. The harvest in a year is usually fifteen animals or fewer (California Department of Fish and Wildlife n.d.). If the RNSP have any role in the management of hunts it is a passive one. The RNSP serve as a de facto elk refuge, and entities responsible for managing hunts can and do use population statistics gathered from elk in the RNSP to assist in making hunt decisions.

Research in national parks might be done to provide a benchmark for comparison to areas more affected by human activities. After all, it is against RNSP policy to allow hunting, timber harvesting as conducted on other lands managed by state and federal agencies, or building of new roads or construction of housing subdivisions and strip malls, all of which are activities that are pervasive in the contemporary landscape of the conterminous United States. Consideration should also be given to activities that occurred in the RNSP before these lands were put under the care of state and federal park services. Meadows are the prime forage habitat for elk along north-coastal California (Franklin, Mossman, & Dole, 1975; Weckerly, 2005). Domestic livestock grazed meadows for decades, if not a century, before those meadows were acquired by the RNSP (Starns, Weckerly, Ricca, & Duarte, 2015). Some parts of meadows were actively farmed for vegetables and fruit orchards (Veirs, 1987). Vegetation that is not native to north-coastal California, such as exotic plants, are present or widespread in many meadows (Hofstra, 1997; Starns et al., 2015). Invariably, dispersal of exotic plants was facilitated by human activities. Unlike now, puma (*Puma concolor*) and black bear (*Ursus americanus*), predators of elk, were hunted and were uncommon throughout the early and middle parts of the nineteenth century (Dasmann, 1964; Orr, 1937). In spite of anthropogenic activities that are pervasive on other lands but not parkland, the human footprint cannot be ignored in the RNSP.

Consequently, I do not view RNSP, either before government stewardship or now, as undisturbed or even "not much disturbed" by humanity. In the RNSP of today, elk-people interactions seem unavoidable despite the RNSP striving to minimize impact on animal population dynamics (Hofstra, 1997). Elk in the RNSP are susceptible to poaching and collide with vehicles regularly; RNSP elk must adjust their behavior to accommodate RNSP visitors, which can be numerous. The most obvious behavioral adjustment is habituation. Many elk in the RNSP ostensibly appear unaware of people until people get within about 10–20 m (Fig. 0.1a, 0.1b). Sometimes elk can be dangerous. In late spring and early summer when people approach recently born young, the mother might charge, and sometimes rear up and lash out with her front legs (National Park Service n.d.). I have been charged by female elk in January when I got too close. Of course, adult males in the autumn mating season can also be dangerous and unpredictable. These conditions suggest a somewhat different or perhaps unique environmental setting that does not conjure to me an environment unimpacted by humanity.

A somewhat unique environment is part of the motivation for studying elk in the RNSP. There might be unexpected findings. No two landscapes are identical: there will always be abiotic and biotic differences. Moreover, a seemingly enormous range of possible ways to affect populations seems plausible when considering life-history variation, the kinds and abundances of predators, behavioral interactions among elk, climate, food supplies, and how forage plants respond to herbivory (Owen-Smith, 2010; Street, Rodgers, Avgar, & Fryxell, 2015). Because no two landscapes are identical, some findings might be difficult to anticipate. Surprises are possible (Lindenmayer, Likens, Krebs, & Hobbs, 2010). Add to this the need for information to manage and conserve elk in the contemporary environmental setting of the RNSP. Given that some findings about distribution and abundance might be surprising and unexpected, the most relevant information for managing and conserving RNSP elk is going to come from studies conducted in the RNSP.

In terms of the anticipated, there are features of the environment that might affect elk demography and population dynamics in ways that are not appreciated at the present time. Summers and winters are cool but the winters are wet and the summers are much drier. In my view, the seasonal climatic variability is not so much in temperature as it is in moisture. The wet is primarily from rain and secondarily from fog. Consequently, late winter and spring are the times of the year when food is more plentiful and nutritious and summer is when nourishing food is more limited (Gogan & Barrett, 1987; Owen-Smith, 2010; Starns et al., 2015). How this particular seasonality to the food supply affects demography and population dynamics has not been examined in much detail. Montane elk in mid-continental areas, in comparison, experience food

Figures 0.1a. and 0.1b. Park visitors observing elk.

FIGURE 0.2. Old-growth redwoods in the background with four elk foraging in an RNSP meadow. The structures are houses and a maintenance building of California State Parks.

supplies that are neither very nutritious nor abundant in late winter and spring; their summer food is more plentiful and nourishing (Hebblewhite, Merrill, & McDermid, 2008). Perhaps as a consequence of the climate, elk in the RNSP are sedentary. Elk spend the entire year in the same general area. Montane elk, quite often, are migratory. Typically, summer range is in high elevation places on mountains where elk consume green, growing plant tissue, whereas in autumn, elk migrate to winter range. Valley floors usually have a shallower snow depth than on nearby mountains, allowing elk access to feed. But those forage plants are dormant or dead and not very nutritious.

Another reason for studying RNSP elk is public education. Elk are a charismatic mega-fauna. people from all walks of life enjoy viewing elk. When I talk with people about the RNSP, invariably, elk are mentioned in the conversation even when I do not tell them what I do and where I do it. This surprises me somewhat because the signature feature of the RNSP are the spectacular trees (Fig. 0.2). As a result of the almost mystical stature of elk, visitors constantly inquire about the elk populations. Knowledge about elk distribution, abundance, and changes in distribution and abundance over time is useful to informing the public and demonstrating stewardship of a public trust.

Furthermore, a solid education involves nuance (Leopold, 1949). Getting to know elk and the land required for their subsistence requires, at the very least, spending time with elk in the places they occupy. There are people that do spend much time in the RNSP but they are a very small minority. Most people visit the RNSP briefly, usually a few hours to a few days. Providing unbiased and verifiable information about the environment is the realm of science. A role for a wildlife ecologist, such as me, is to study elk that inhabit the RNSP to detect patterns, provide details of their patterns, and understand the reasons for those patterns. In short, my role is to uncover nuances and communicate those nuances. That way, people that live in urban areas, which is the vast majority of humanity in the developed world, can learn more details about the elk they observe in their brief visits to the RNSP.

An education about elk in the RNSP might also be beneficial to managing and conserving elk in other parks along the west coast. Examples include Point Reyes National Seashore in Marin County to the south and Dean Creek Wildlife Area near Reedsport, Oregon, to the north. Long-term study of tule elk at Point Reyes has occurred but there are no long-term studies of Roosevelt elk that I am aware of in state and national parks along the west coast (Cobb, 2010; Howell, Brooks, Semenoff-Irving, & Greene, 2002). My work might be helpful to showing how to establish long-term studies of elk populations on the cheap, and to assessing the reliability of population statistics and what can be obtained from this information for purposes of managing and conserving elk in park settings.

Preservation and persistence of native animal life is clearly one of the purposes of national as well as state parks (Leopold, Cain, Cottam, Gabrielson, & Kimball, 1963). Prairie Creek Redwoods State Park, now part of Redwood National and State Parks, was created, in part, as an elk reserve (Harper, Harn, Bentley, & Yocom, 1967). In 1933 Prairie Creek Redwoods State Park was expanded to encompass elk forage habitat. Given the clear mandate for biotic preservation, elk can be assumed to inhabit the parks in perpetuity (Leopold et al., 1963). To assess the likelihood that elk will persist in the RNSP in perpetuity is truly impossible. The fossil record is clear: climate and biota are constantly changing. Persistence is addressed in the sense of how elk persevere in the RNSP during the time of my study. What are the factors and processes that are driving distribution and abundance of elk populations? As long as the environmental factors that affect distribution and abundance during the time of my study exist into the future, my conclusions about elk persistence should hold. Furthermore, if environmental factors do change in the future, then my findings should be useful to assessing the effect of changes on elk distribution and abundance.

Persistence is not used in the sense as it is used in the management of threatened and endangered species. That sense is about what it might take for elk to persist to some future point in time with a predetermined probability. Modeling "what it might take" persistence is the domain of population viability analysis (Duarte, 2015). If I were to take this route, one hypothetical outcome is as follows. A population viability analysis might indicate that all available elk habitat is saturated, yet more elk will be needed to ensure, with 95 percent probability, persistence 50 years into the future. A means of increasing elk abundance could be increasing elk forage habitat, which in the RNSP is primarily meadows. This remedy is unrealistic because it would entail removing forests containing coastal redwoods (*Sequoia sempervirens*). Since the RNSP have the largest existing stands of old-growth coastal redwoods and most of the RNSP were created to preserve coastal forest, this option is untenable. Because elk in the RNSP are listed neither as threatened nor endangered on a state or federal register, managing elk as if they are scarce or soon to be scarce is impractical. The resources are simply not available to manage elk in the RNSP as if they are threatened or endangered.

Nonetheless, elk persistence can be evaluated using concepts that are actively employed in the science of scarcity, resiliency and redundancy. Resiliency refers to how resistant a population is to disturbances or factors that negatively affect abundance. Resiliency can be evaluated by the abundance of the population, demographic processes such as recruitment of young and adult survival that govern its change, and the abiotic and biotic factors influencing demography. I assume that the more abundant the population, the higher the resilience. More animals indicate a stronger possibility of not decreasing to the point of extinction. The name of the game, however, is to estimate what is or can be the most abundant, sustainable population. A sustainable population then has to be the most abundant population that does not graze the food supply to an extent where the productivity of plant communities is degraded or plants become less palatable or more resistant to herbivory (Ricca, 2013). In other words, a population that has arrived at an equilibrium with its food supply or the largest population size given the level of offtake from agents of mortality such as predators, disease, hunting, vehicle collisions, and poaching.

Redundancy is greater when there are more populations in a landscape. If one or a few of the populations in a landscape go extinct, the species will persist because of the remaining, extant populations. Redundancy then requires establishing spatial boundaries of populations and understanding something about what maintains those boundaries so the populations in the landscape can be counted. Establishing the spatial boundary of a population might be easy. The boundary might simply be the spatial extent of habitat that provides food and cover for the species. It might also be more complicated, such as when two or more populations partition the same habitat patch when, seemingly, there are no discontinuities in resource distribution to explain the occurrence. In such a setting, social behavior might be involved (Hestbeck, 1982). Habitat partitioning usually involves some degree of overlap between populations, yet the populations rarely inhabit the same place at the same time (Julian, Schmidt, & Weckerly, 2013; Kolbe & Weckerly, 2015).

Habitat partitioning by adjacent populations raises the possibility of an interesting dilemma about redundancy and resilience. If, for example, populations that partition habitat fuse into one population, then the new population will be more abundant, increasing resilience but reducing redundancy. On the other hand, what about one population splitting into two populations? Is elk persistence in a landscape more likely when there is greater redundancy or higher resilience? Yet, given the poor understanding of what prompts habitat partitioning by elk populations it is unclear how frequently population fusion and fission occurs, and thus the dynamics between redundancy and resiliency are also unclear.

The reasons laid out essentially boil down to understanding something about the forces driving distribution and abundance, a fundamental goal in ecology and part and parcel of any management program. "Management without information would be a dangerous policy indeed" (Leopold et al., 1963). To understand something about the drivers of distribution and abundance is going to require a framework to guide my research activities—models. The selection or development of models is a serious matter. To tweak a famous saying by G. E. P. Box, Hunter, & Hunter (1978, 440): "all models are wrong but hopefully my models are useful." My models will be too simple, but hopefully they possess the essential processes and structure to help show me where to look for patterns and shed some light on the patterns I detect in data. Models can be mathematical or expressed in words and diagrams. Since I am a poor mathematician, my models are in words and at times facilitated by diagrams and, paradoxically, formulas. Because I am a poor mathematician the formulas I present are rather straightforward and thus should be helpful to understanding models and diagrams.

The first models presented are about distribution, what dictates the places elk use. Individual elk tend to consume and use similar kinds of resources but often occur in different places. Elk that tend to occur in the same places also interact with one another. Elk in the same place at the same time is how I define a population (Ehrlich & Daily, 1993). The models after distribution are about population dynamics. These models allow me to assess the factors that drive increases and decreases in population abundance over the years. With some understanding of population drivers, it is then possible to estimate the population size that can be sustained by the environment.

DISTRIBUTION

Partitioning of habitat, or spatial boundaries between adjacent populations that seemingly have no discontinuities in elk habitat, does occur in the RNSP (Julian et al., 2013; Kolbe & Weckerly, 2015). Forces that probably have considerable influence on resource partitioning are food distribution and predation. Ideal free, not ideal despotic, distribution appears insightful to understanding elk distribution across a landscape (McLoughlin, Boyce, Coulson, & Clutton-Brock, 2006). Theoretically, ideal free distribution occurs when animals are free to choose habitat that optimizes fitness (survival, reproduction, and recruitment of young). Animals in populations should select habitat that optimizes food intake and reduces vulnerability from predators. In the real world of the RNSP, fences brought about by social behavior of the elk themselves would seem to prevent the free movement of elk. Ecological factors that govern availability of food supplies to individual elk should, therefore, be similar among populations. The availability of food might be more restricted than the actual biomass of forage. Such a setting might occur when lethal risks, such as being killed by a predator, are high (Bowyer, Kie, & Van Ballengerghe, 1998). A convenient way to view food availability is a risk:food ratio. When actual forage is limited, populations should partition habitat such that members of each population have similar risks from lethal factors in relation to food supplies (Heard, 1992; Street, Weckerly, & Schwinning, 2013).

I assume elk are less vulnerable but not immune to being killed by natural predators in meadows. The most effective predators of elk in the RNSP are puma and black bear. Puma hunt alone and stalk their prey. Puma will kill juvenile and adult elk whereas black bear appear to prey most heavily on neonatal elk (Elbroch, Lendrum, Newby, Quigley, & Craighead, 2013; Griffin et al., 2011). Black bear presumably search for neonatal elk that bed in cover and remain motionless or hide. Because of the large body size of elk, puma and black bear seem to have the greatest chance of killing elk when they surprise animals. Surprise would be helpful to black bear by reducing the prospect of mothers defending young (Harn, 1958). The element of surprise should be more likely in broken terrain with dense vegetation than in flat, open habitat (Atwood, Gese, & Kunkel 2009; Husseman et al., 2013).

When forage habitat is where animals are less vulnerable to lethal risks and food distribution and abundance is predictable or stable, there should also be predictable patterns between food abundance and elk use (Riotte-Lambert, Benhamou, & Chamaille-Jammes, 2015; Van Moorter, Visscher, Benhamou, & Borger, 2009). Since elk populations in the RNSP can show tenacity to particular meadow complexes, it is likely that elk have the capacity for long- and short-term memory about resource distribution (Berger-Tal & Bar-David, 2015; Franklin et al.; Harper et al., 1967). Long-term memory would help elk realize the overall patterns of resource distribution and abundance and short-term memory would be helpful to elk deciding when to leave a forage patch. When forage distribution is stable at the scale of weeks to months, the percentage use of patches by elk should be the same as the percentage of the food supply available in patches. For simplicity, I assume that forage habitat is continuous or that discontinuity imposed by non-forage habitat is not expansive enough to require hours of travel by elk to get through.

Social interactions outside of the mating season might also have a role in maintaining boundaries between populations. Aggressive interactions among population members are such that they occur quickly and, in most instances, without physical contact (Franklin & Lieb, 1979; Weckerly, 1999). Conflicts between members of a population are inevitable (McCullough, 1969). Population members have a social hierarchy and settle conflicts with ritualized behaviors that have a small chance of the two elk actually touching one another. Young animals receive the brunt of aggressions (Franklin & Lieb, 1979), yet young should remain in their natal populations because the population members provide increased detection of and dilution from attack by predators. Young should also benefit from becoming aware of the distribution and abundance of resources within and across seasons (Rutberg, 1986; Sueur et al., 2011; Weckerly, 1999).

I mention social interactions outside the mating season because of adult males. Adult males during the mating season often engage in aggressive interactions that involve physical contact that can cause physical injury (Geist, 1986). The aggression, furthermore, probably has little consequence to maintaining spatial boundaries between populations because the intent of male aggression is to mate. Therefore, males might seek females in different populations to increase mating opportunities.

Elk should be less familiar with members of other populations. Elk that are unfamiliar with one another are more likely to engage in behaviors that escalate into aggression involving physical contact, which increases the potential for injury. When adjacent populations are in close proximity or fuse together, members of different populations probably engage in aggression that escalates in intensity (McCullough, 1969; Thouless & Guinness, 1986; Weckerly, 1999). Because escalated aggression probably has few rewards to forage acquisition, in regard to the risk of physical injury, members of each population might be vigilant to one another. Prolonged vigilance might also have detrimental consequences to food intake simply because animals with the head up cannot graze. In the end, populations go their separate directions—away from one another. Social behavior is a day-to-day mechanism that helps sort out population distribution so that, over a lifetime, members within and among populations experience similar food supplies and risks from lethal factors (McCullough, 1979).

The environment in and around the RNSP is not static. Population boundaries can change when lethal risks become elevated or food distribution suddenly changes. If poaching, for example, becomes more frequent on two adjacent populations, then those populations might come together and fuse into one. The potential costs to foraging and the risks of physical injury from escalated aggressions are outweighed by the reduction in lethal risk. On the other hand, what if food suddenly becomes more abundant? In the contemporary setting around the RNSP, cattle and the barriers that are intended to keep elk out, such as fences, are sometimes removed. In such a setting, there should be colonization by nearby elk. If there are two or more populations in the vicinity, these populations should come together in the new area in response to risks posed by an unfamiliar place (Street et al., 2013; Wolf, Frair, Merrill, & Turchin, 2009). An inevitable outcome will be increased aggression between unfamiliar elk. The costs of aggression from adjacent populations fusing, in this setting, are outweighed by the abundant, new food source. Furthermore, familiarity among animals ought to increase over time. The result should be the emergence of a new social hierarchy where aggression that escalates to physical contact declines (Hirotani, 1990).

In contemporary settings, forage habitat such as pastures also can become inaccessible in a very short time. If a fence that at one time was a barrier to elk becomes riddled with holes or closed gates are opened, then elk have ready access to the pasture. If the holes in the fence are quickly repaired and the gates closed, each elk in the population should then experience a sudden reduction in the food supply. As such, increased conflicts over the remaining food and escalated aggression should be the result. One short-term response by the population might be to lessen the opportunity for escalated aggression (Weckerly, 1999). Alleviating frequent aggression could be achieved by the population splitting into smaller groups or subpopulations. Each of these smaller subpopulations would consist of elk that are more familiar with each other or that become familiar quickly. Animals that are familiar with each other will probably engage in frequent aggression, but the intensity of the aggression should be low (Thouless & Guinness, 1986). If less time is spent engaged in or avoiding the opportunity for escalated aggression, perhaps more time can be spent memorizing food distribution. Memorizing the distribution of food seems to enable efficient exploitation of forage in competitive settings when patch locations remain stationary across months and forage plants grow after being grazed (Riotte-Lambert et al., 2015). Each of these subpopulations becomes a new population when the area of use overlaps little with other populations. Spatial separation should reduce encounters between populations and increase forage efficiency.

POPULATION DYNAMICS

I define a population as a female herd. Females spend much of their time, particularly in winter and early spring, aggregated with the same females, a socially bonded herd (Franklin & Lieb, 1979; Franklin et al., 1975; Weckerly, 1999). Because there is temporal and spatial variability in the extent of social bonding, or fission–fusion dynamics, herds can comprise two or more subherds or groups (Franklin & Lieb, 1979; Sueur et al., 2011; Weckerly, 1999). The herd splits into groups that later on fuse into one herd or a small number of groups. Few adult males occur in or remain for long in herds except during the mating season (Weckerly, Ricca, & Meyer, 2001). Herds of socially bonded females are spatially separated from other such herds which might, furthermore, have consequences for fine-scale genetic substructure (Coltman, Pilkington, & Pemberton, 2003). The recruitment of young into these herds seems to be more strongly affected by the number of females in the herd and not the number of adult males that are in the nearby area or the few that might be associated with the herd (Julian et al., 2013; McCullough, 1979).

A definition of a population that does not include males is a tad odd. In a sexual species like elk, males and females are required for population persistence through reproduction. The number of males required to mate with females is the issue. Elk have a polygynous mating system (McCullough, 1969). Males attempt to mate with multiple females and there is considerable variability among males in the number of copulations (Clutton-Brock, Rose, & Guinness, 1997). Across years there is also variation in the adult sex ratio (Weaver & Weckerly, 2011). Variation in the adult sex ratio does not appear to influence rates and timing of conception until sex ratios are highly skewed toward females and the age structure of sexually mature males is young (Noyes, Johnson, Bryant, Findholt, & Thomas, 1996). Furthermore, intersexual resource partitioning is common in large herbivores and usually manifests as males having little influence on female demography (Bowyer, 2004). Males also tend to wander across the landscape because males display lower site tenacity than females at the scale of months (Weckerly, 1993). Polygyny, reproduction, and intersexual differences in resource use and site tenacity complicate how to define a population. One possible alternative is to delineate superpopulations (Weckerly, 1993; Williams, Nichols, & Conroy, 2002). A superpopulation is all the animals, female and male, that use an area. Owing to the inherent complexities of a superpopulation I stick with populations. I chose to define a population thinking about practical and operational considerations. I needed population boundaries that could be measured in the field and population dynamics that had the potential to differ from other populations. The spatial boundaries of female herds have been delineated and display dynamics that are unaffected by the changes in abundance of other female herds, even those close by (Julian et al., 2013). Uncoupled herd dynamics suggest there is little inter-herd movement by individual females. Each female herd was a population.

Because herd dynamics were uncoupled from one another, I was less interested in a larger metapopulation (Levins, 1969). By definition, a metapopulation is a collection of spatially structured populations in a heterogeneous landscape. Metapopulation analysis involves understanding not only the dynamics of individual populations, but also the population processes that influence the larger metapopulation (Hale, Treml, & Swearer, 2015; Lee, 2015). Because each population seems to behave independently, I focused on individual populations more than on all of the populations. There also is the element of partitioning the same habitat by multiple populations that is usually not considered in metapopulation analysis. The most interesting aspects about the population ecology of RNSP elk seemed to be at the level of the population, more than at the next level up, the metapopulation.

There are two parts to the study of population dynamics. The first part is measuring population abundance over a series of years so that population dynamics can be described. Abundance estimates measured annually over a number of years are collectively called a time series. For now, I forgo information on how I measured abundance and some of the demographic variables. In Chapter 2 I present my tools and why I used the particular tools that I did. I view the description of population dynamics as primarily methodological; although it was personally rewarding to develop my population survey methods, it was not the primary motivation for the study.

The second part to the study of population dynamics is understanding the drivers of population change. In the broadest sense, population change is brought about by fluctuations in inputs (births and immigrations) and outputs (deaths and emigrations) (Duarte, Weckerly, Schaub, & Hatfield, 2016). The roles of inputs and outputs to population dynamics are estimated in the form of relationships that are density dependent or density independent, and in the interactions between density-dependent and density-independent factors (McCullough, 1979; Starns, Ricca, Duarte, & Weckerly, 2014).

Density dependence in populations of large herbivores is pervasive (Brook & Bradshaw, 2006). There are two simple, conceptual descriptions that are often used to frame the outcomes of a competitive environment to sustainable population growth, or population regulation: (1) the standard-logistic and (2) the theta-logistic models (Fig. 0.3). The standard-logistic has been around for some time and the theta-logistic is a more recent development (Clark, Brook, Delean, Akçakaya, & Bradshaw, 2010; Gilpin & Ayala, 1973). Both models describe a small population that grows rapidly because of an abundant food supply. When the food supply for herbivores becomes more limited owing to consumption by more mouths, population growth slows and eventually the population comes into equilibrium with its food supply. It should be noted that neither form of the logistic model explicitly modeled consequences of logistic population growth to the food supply (Caughley, 1976).

The two forms of the logistic model differ in one parameter: the shape parameter theta (θ). Theta is probably greater than 1.0 for larger herbivores (Boyce, 1989; Fowler, 1981; McCullough, 1979). There are two subtle distinctions to population growth between the two forms of the logistic model. These distinctions can be illuminated by paying attention to what occurs between the inflection point (I) and when the population is in equilibrium with its food supply (K). The inflection point is the abundance with the maximum number of recruits (product of recruitment rate and abundance). For the theta-logistic

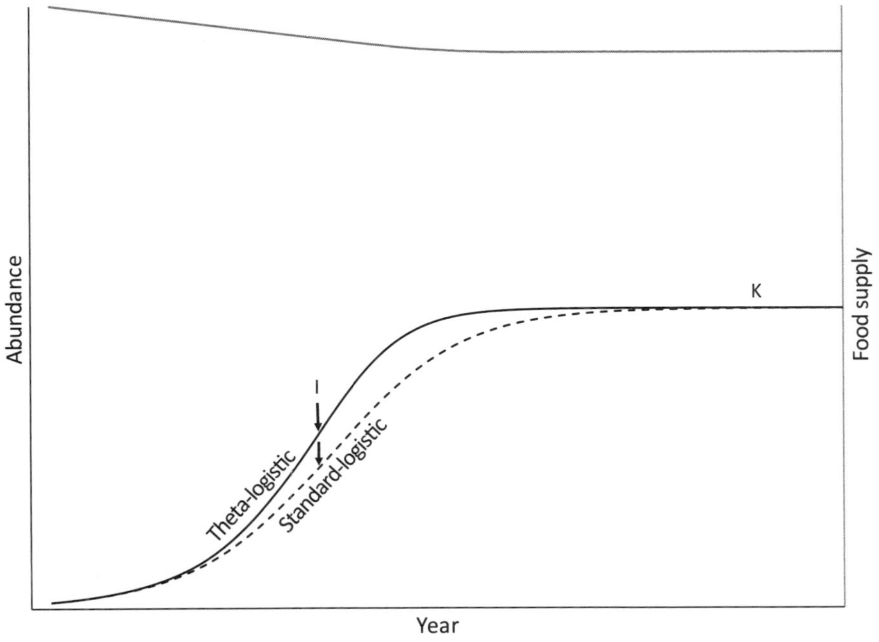

FIGURE 0.3. Standard- and theta-logistic population growth across years. A general, conceptual response of the food supply to herbivore population growth is also presented (Caughley 1976). The population begins with two animals. The inflection point (I) is when the maximum number of young are recruited into the population; K denotes when the population is in equilibrium with the food supply, referred to as K carrying capacity. The formulas for the discrete form of the standard-logistic model are $N_{t+1} = N_t e^{\left(r\left(1-\left(\frac{N_t}{K}\right)\right)\right)}$ and $N_{t+1} = N_t e^{\left(r\left(1-\left(\frac{N_t}{K}\right)^{\theta}\right)\right)}$ for the theta-logistic. Parameter values were = 0.28, K = 100, and θ = 1.5. The parameter r is often referred to as r_{max}, the maximum rate of intrinsic growth ($ln\left(\frac{N_{t+1}}{N_t}\right)$) in the absence of density dependence or when abundance is very small. Abundance in a year is N_t and abundance the year after is N_{t+1}.

model, the abundance at I is closer to K, whereas for the standard-logistic model the abundance for I is precisely one-half of K. An outcome of this subtle difference is that the inverse relationship between abundance and intrinsic rate of increase is linear for the standard-logistic and nonlinear but still inverse for the theta-logistic (Fig. 0.4). The theta-logistic model has captured population growth of large mammals (Boyce, 1989; Fowler, 1981; McCullough, 1979). Theta-logistic population growth makes sense for large herbivores (Fowler, 1981). Large herbivores are long-lived and must, therefore, have a high survival rate. A high survival rate implies that large herbivore populations probably spend much time between I and K. Over the long term, life history characteristics are likely to be shaped by a competitive environment (Bowyer, Bleich, Stewart,

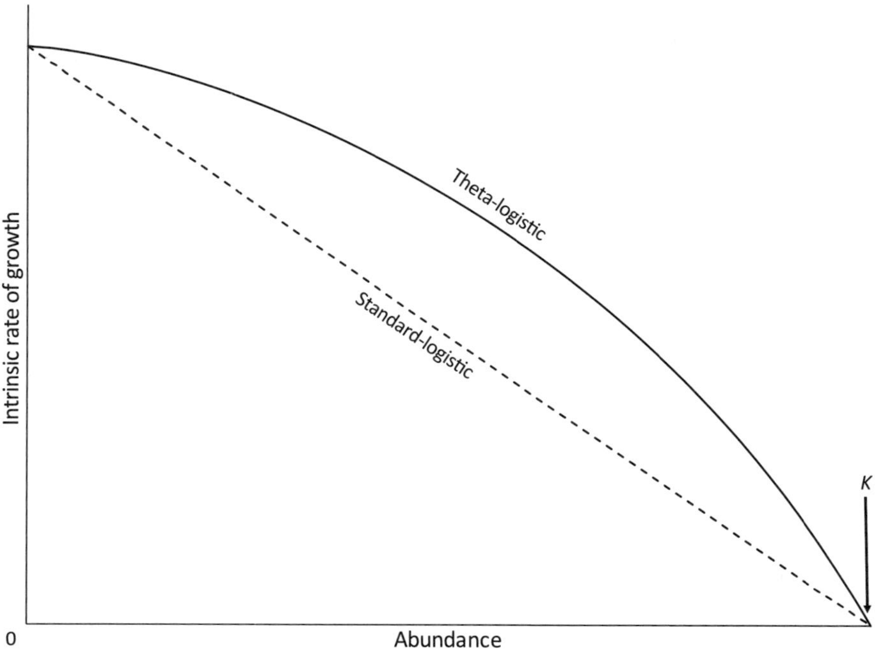

FIGURE 0.4. Predicted relationships between abundance and intrinsic rate of growth for the standard- and theta-logistic models. Note that when the population reaches K the intrinsic rate of population increase is zero.

Whiting, & Monteith, 2014; McCullough, 1979). The potential outcome to a competitive environment among individuals over evolutionary time is population growth that displays a high rate of increase across a wide range of abundances from very low to I. Be that as it may, there are also examples of growth of free-ranging populations that are consistent with the standard-logistic model (Street et al., 2015; Stewart, Bowyer, Dick, Johnson, & Kie, 2005).

A disquieting issue with the theta-logistic model is the additional parameter, θ. Theta should be a value unconnected to the value of r. Yet, these parameters can be correlated to one another when the theta-logistic model is estimated from abundance data (Clark et al., 2010). Why estimate two parameters when one tells you something about the other? The issue of dependent parameters seems to be most acute when there is a time series of abundances that are between I and K or around K (Polansky, de Valpine, Lloyd-Smith, & Getz, 2009). In that case, there are not abundances that span the range from very low to K, which is crucial to the estimation of θ. By the way, most of my time series were gathered when populations were probably above I. In such a setting, estimated relationships to summarize population growth might be linear, and so might favor the standard-logistic model, simply because of the narrower range of abundances. Because of this issue with estimating the one additional parameter, θ, I rely on the standard-logistic model.

Because I viewed large herbivore populations as spending much of the time above the inflection point (I), animals are most likely in a competitive setting. In a highly competitive environment, adult females face a trade-off: allocate energy and nutrients to themselves or allocate them to provision offspring (Barboza, Parker, & Hume, 2009). Conversely, in the less common instance where food is abundant or individuals are better competitors, females in good condition will have adequate protein and fat stores to provision a developing offspring during gestation and also provide milk during lactation. Lactation is the most energetically demanding time in the life of a female (Barboza et al., 2009). This means that in all likelihood the number of births fluctuates to some degree across years, and juvenile survival tends to be lower and more variable than adult survival (Gaillard, Festa-Bianchet, & Yoccoz, 1998).

Another model of population growth that has been applied to a variety of long-lived species is the Gompertz model (Dennis, Ponciano, Lele, Taper, & Staples, 2006; Hatfield, Reynolds, Seavy, & Krause, 2012; Seavy, Reynolds, Link, & Hatfield, 2009; Taber & Gogan, 2002; Wang et al., 2009). The Gompertz model is a somewhat different beast from the standard-logistic model in that the Gompertz model requires natural log transformations of abundance in the present and the previous year (Box 0.1).

Appeals of the GSS are that K carrying capacity can be estimated and the state-space model has been developed to disentangle a time series of abundances into the processes that actually drive population change (i.e., birth, death, individual ingress or egress) called process variance from distortions in abundances due to the procedures used to count populations. Distortions in abundance estimates due to sampling procedures are called observer error. Observer error, by the way, occurs in most wildlife population surveys.

In the GSS model, carrying capacity is a long-term average of a stationary distribution. As such, periodic oscillations from such factors as time lag influences on demography are not part of the concept of a stationary distribution (Dennis & Costantino, 1988). The influence of time lags on population dynamics has a long history of study in population biology. Nonetheless, the estimate of carrying capacity from the GSS model can been viewed as K carrying capacity (Hatfield et al., 2012; Saether et al., 2008).

My attention to density dependence does not mean that unpredictable, density-independent factors are rare or uninfluential. There are many examples of large herbivores succumbing to harsh winters with much snow and ice

Details about the Gompertz model

Density-dependent growth in the Gompertz model is estimated with a linear, first-order autoregressive time series model:

$$X_t = a + bX_{t-1} + E_t. \qquad 0.1$$

In this model X_t is the natural logarithm of true abundance in a year and X_{t-1} is the natural logarithm of true abundance the previous year; *a*, *c* (1 + *b*), and *E* represent population growth in the absence of density dependence, strength of density dependence, and process variance, respectively. When *c* is less than 1.0 density dependence is affecting population dynamics. Moreover, the lower the *c*, the stronger the density dependence. Process variance is a measure of population change due to demography and animals that move into or out of populations. Procedures used in the field to collect abundance data usually distort the actual abundance by either under- or over-counting the actual abundance. The distortion can span a spectrum from a little to a lot. Observer error is modeled in a Gompertz State-space (GSS) model as

$$Y_t = X_t + F_t \qquad 0.2$$

where F_t is observer error that has a Gaussian or normal distribution. I used the Gompertz model to predict changes in herd abundances or states: in other words, a GSS model.

or to droughts in arid regions (Marshal, Cain, Bleich, & Rosenstock, 2009; Tyler, 2010). Usually these density-independent or climatic factors operate through restricting the food supply. Forage plants are unavailable because of deep snow and ice or are rendered difficult to digest by drought or cold winter temperatures.

Yet, the intensity of density-independent factors can be mediated by abundance of the population. When population abundance is high, food is more limited, and animal condition is often poor, animals are then more vulnerable to density-independent factors that affect the settings that dictate growth and abundance of forage plants. Such settings are often harsh winters, droughts, or disturbances such as floods and fire. At the other end, when population abundance is low, animals have protein and fat stores that make them less vulnerable to the vagaries of climate and disturbances. Indeed, there can be interactions between density-dependent and density-independent factors (Saether, 1997).

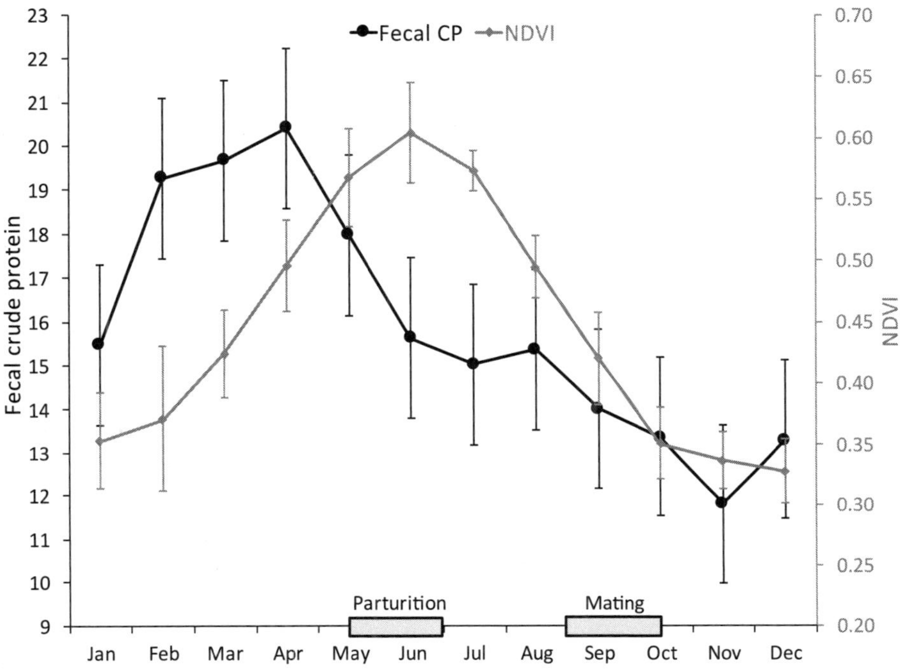

FIGURE 0.5. A line graph, with error bars, to assist in explaining the potential consequences to elk recruitment from the environmental setting in the RNSP. Displayed is the mean food supply (NDVI) and mean dietary nutrition (fecal CP) for each month of the year. Also presented are seasons of birth (parturition) and mating. Fecal crude protein (CP) is an index of dietary nutrition. Higher values indicate a more digestible diet. Dietary nutrition is greater during plant growth in late winter into spring than at the peak of the food supply, and in summer and autumn when forage plants are dormant, slow growing, senescent, or dead. The biomass of the food supply in forage habitat is indexed from an NDVI. Higher NDVI indicates more forage. The error bars are 95 percent CI and reflect the variability in means. Fecal CP data were collected from the Boyes and Gold Bluffs herds, and the NDVI measurements were obtained from Boyes Meadows (see Chapter 2). Data were gleaned from Gogan & Barrett (1987), Starns et al. (2015), and Harper et al. (1967).

The seasonal phenology of the food supply can also influence demography and population dynamics (Fig. 0.5). Forage in the RNSP is probably most abundant and nutritive in late winter and early spring, which is before young are born and mothers are burdened with the enormous, energetic demands of lactation. During late winter and early spring, forage plants are growing rapidly. The ratio of plant cell wall content to cell soluble material is probably low. Much of the cell soluble material is easily digested by large herbivores, and many of the cell wall constituents are also digestible. Digestion of cell wall constituents, however, requires fermentation that can take hours, but usually requires a day or more (Barboza et al., 2009). During the time when lactation occurs, forage is abundant but the nutritive value of forage is probably declining. Because of the demands of lactation, protein and fat stores are probably reduced or, perhaps, depleted. Offspring are usually weaned from their dependence on mothers' milk by about 90–120 days after birth (Kristan, 1992, Pollard, Asher, & Littlejohn, 2002). Offspring will often continue to suckle after 120 days, but the frequency is much less than shortly after birth. At about 90 days of age the digestive system is also developed to an extent that young can survive and grow when consuming plant material (Hammond et al., 2013). After weaning offspring, mothers then have the mating season that occurs in late summer and early autumn. The probability of whether an adult female conceives is positively related to her condition (Cook et al., 2013). The greater her fat and protein stores, the greater the chance she becomes pregnant (Barboza et al., 2009). Conception, parturition, and lactation might be affected by late winter and early spring food supplies and not just by food supplies in summer and autumn. Recruitment of young through birth should then be dictated by environmental conditions before mothers conceive.

Since reproduction is a risky venture to females, reproductive pause is not unusual in a long-lived species like elk. Female elk can live 15 or more years (McCullough, 1969). The first mating season after weaning young, females might not be in sufficient condition to meet the energetic and protein demands of provisioning young and, consequently, might delay reproduction for another 1 or 2 years (Stewart et al., 2005). Considering the seasonal phenology of the food supply, there is probably a lag of 1 or more years to density-dependent and density-independent relationships (Starns et al., 2014).

I spent 20 years collecting data. Twenty years were needed to delineate population boundaries and describe population dynamics and to gather data to estimate processes influencing population change. From the beginning, models for exploring population dynamics were in place because the density-dependent and density-independent paradigm was well established (McCullough, 1979; Saether, 1997). The seeds of the distribution model were sown from early studies (Weckerly, 1998, 1999, 2001; Weckerly et al., 2001). As patterns began to emerge in the study, I conducted much of the work to test the distribution model. In the course of collecting these data, I got lucky: two adjacent herds came together at the same place at the same time. It was not until the very end of the study that I learned how to use the GSS model (Dennis et al., 2006; Duarte et al., 2016; Hatfield et al., 2012; Seavy et al., 2009).

Following this introduction are chapters on the history of elk in the RNSP; techniques and my analysis of why I used count data to estimate abundance; and my findings on distribution, abundance, and their drivers, including unexpected findings. This information was then brought together to assess redundancy, resiliency, and management. As is somewhat evident in this Introduction, I not only used the data I, Mark Ricca, Richard Enderlein, Curtis Miller, my folks, RNSP biologists, and students collected, but I also used data collected from previous studies with elk in the RNSP and elsewhere. My shoestring study could not have been done without two things. First, the willing, able, and gracious assistance of others; and second, the information I gleaned from the literature.

1

Elk History and Notes on Large Predators

The remaining band of any size of Roosevelt elk occupy a strip twenty miles long by four miles wide near the ocean just north and west of Orick.

— Earl Barnes (1925a), Game Warden, Eureka, California

In ecological studies, what has gone on in the past can dictate what is observed in the present. Which leads me to thoughts on how far back in history I should go? Since the study is about elk, the beginning is the origin of the species.

Cervus elaphus (i.e., North American elk) probably originated during the late Miocene and early Pliocene (about 5–7 million years ago) in central Asia between what is now Kyrgyzstan and northern India (Lorenzini & Garofalo, 2015; Ludt, Schroeder, Rottmann, & Kuehn, 2004). This area is somewhere between 32 and 42°N latitude. At those latitudes, a predictable, seasonal environment was likely. Food in winter was probably less nutritious because of cool or cold temperatures and summer food was probably more plentiful and nourishing because of warmer temperatures. At about the same time as the origin of *Cervus elaphus*, global climate change occurred and grassland habitat also became more apparent. This is a noteworthy detail because grass is a large dietary component of contemporary Roosevelt elk (Harper et al., 1967; Jenkins & Starkey, 1991).

From central Asia two separate lineages of *Cervus elaphus* emerged. One line dispersed west to Europe, North Africa, and Mediterranean areas; and the other lineage spread east to eastern Siberia and North America (Lorenzini & Garofalo, 2015; Ludt et al., 2004). The archaeological and genetic evidence indicates that elk colonized North America by crossing the Bering land bridge between Siberia and North America during a period of global cooling and expansion of glaciers about 15,000 years ago. The colder climate at the time resulted in a drop of sea levels (Meiri et al., 2014). The terrestrial communities on the land bridge and parts of Alaska were diverse and productive, and elk were probably attracted to the food supply. Elk then proceeded to colonize areas to the south in North America as glaciers receded. Around the same time that elk crossed the land bridge and began colonizing the Americas, so did humans (Guthrie, 1967). Elk and humans appear to have a history of interactions.

In the present day, *Cervus elaphus* in Europe are called red deer and in eastern Asia and North America elk or wapiti. I call them elk simply because it is a widely used label in North America (Weckerly, 1996). Species distinction for each form seems warranted given the disjunctive geographic ranges as well as morphological and behavioral differences. Obvious differences are the larger body size of elk and the tan or browner coat color than the namesake of red deer, the reddish pelage coloration. Furthermore, during the mating season red deer males roar, a guttural sound, whereas male elk bugle, which sounds like a whistle to me. Species distinction for red deer in Europe, North Africa, and Mediterranean areas versus elk in North America and eastern Siberia has also been advocated because of the east–west radiation from a common ancestor in central Asia (Lorenzini & Garofalo, 2015). Nonetheless, when the two forms have been purposefully brought together, red deer and elk do not have reproductive barriers; they are sexually compatible and the offspring from red deer–elk mating develop into adults that can also reproduce (McCullough, 1969). For the sake of simplicity, I treat red deer and elk as one species and hereafter refer to them as elk.

Roosevelt elk (*Cervus elaphus roosevelti*) are the subspecies that occur in the RNSP. The RNSP are near the southern extent of their contemporary geographic range. Mendocino County, the next county to the south of the RNSP, also has Roosevelt elk (Meredith et al., 2007). McCullough (1969) reported that in the 1800s Roosevelt elk range extended farther south into the county below Mendocino County: Sonoma County (Fig. 1.1). From the southern end of where Roosevelt elk occurred, the range extended north to Vancouver Island and southern British Columbia. Between this north–south extent Roosevelt elk were historically considered to occupy areas eastward to the western slope of the Cascade Mountains and the Marble and Trinity Mountains in northern California (McCullough, 1969). Market and unregulated hunting extirpated elk from much of this range in the late 1800s and early 1900s. In the present day, the eastern extent of Roosevelt elk range in Washington and Oregon is

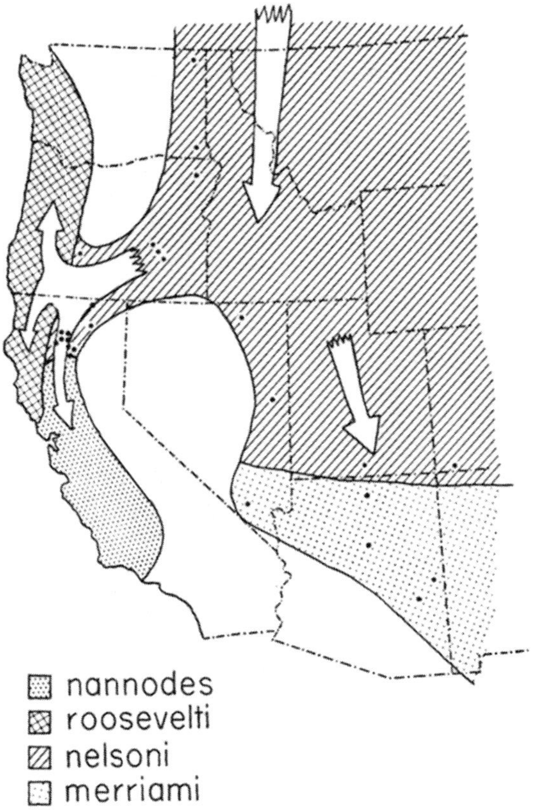

FIGURE 1.1. Hypothesized route that elk colonized California (McCullough 1969, Figure 2). The Californian subspecies are Rocky Mountain elk (*Cervus elaphus nelsoni*), tule elk (*Cervus elaphus nelsoni*), and Roosevelt elk (*Cervus elaphus roosevelti*). Some of the geographic range of the extirpated Merriam's elk (*Cervus elaphus merriami*) is also provided on the map. The dots are localities of specimens examined. These specimens were used to delineate geographic boundaries of the subspecies.

arbitrarily set at Interstate 5 (Cook et al., 2013; Meredith et al., 2007). In the past century, relocation efforts to repopulate the historic range sometimes used Rocky Mountain elk (*Cervus elaphus nelsoni*). Now there are locations on the western slopes of the Cascade Mountains that have Rocky Mountain elk and there are other localities where Roosevelt elk and Rocky Mountain elk have intergraded.

I should also mention where Roosevelt elk were purposefully introduced and became established outside of this geographic range. These introductions were on islands off the coast of southeastern Alaska. Elk were probably present in Alaska from the time they crossed the Bering land bridge to as recently as 3,000 years ago (Guthrie, 1990). Yet, Roosevelt elk are not considered endemic to Alaska. To my knowledge, Roosevelt elk inhabit Afognak and Raspberry Islands (Maier, 2001). Elk have established themselves on Etolin Island after introductions of Roosevelt elk and Rocky Mountain elk (Kirchhoff & Larsen, 1998). Besides Roosevelt elk along the north coast of California, two other subspecies of elk are recognized in the state (Fig. 1.1). To the south, tule elk (*Cervus elaphus nannodes*) and to the east, Rocky Mountain elk. Tule elk tend to be smaller in body size and can be somewhat lighter in coat color than the other two subspecies. Roosevelt elk tend to be darker in coat color and the largest in body size of the three subspecies. Roosevelt elk do not necessarily possess the largest antlers, however (McCullough, 1969). That distinction goes to the Rocky Mountain elk. Rocky Mountain males with large antlers have longer main beams and tines than large-antlered Roosevelt and tule males.

McCullough (1969) hypothesized the geographical route of elk colonization into California (Fig. 1.1). After crossing the Bering land bridge, elk dispersed south to mid- and east-continental areas; some dispersers headed to the west coast near the present-day Oregon–California border. Once animals arrived at the coast, elk colonization proceeded north. If this route of colonization is correct, then Roosevelt elk have inhabited the RNSP earlier than more northern areas of their geographic range. The direction of movement is based on the southerly direction of dispersal after crossing the Bering land bridge and morphological differences among the three subspecies in California (Guthrie, 1967; McCullough, 1969). Tule and Roosevelt elk are considered offshoots from the morphologically generalized Rocky Mountain elk. It is unlikely that tule elk or Roosevelt elk derived from the other because body size has evolved in opposite directions; Roosevelt elk are the largest, tule elk the smallest, and Rocky Mountain elk are intermediate. The body size patterns of the three subspecies concur with genetic analyses conducted on present-day Californian elk (Meredith et al., 2007). Tule elk and Roosevelt elk display the greatest genetic differences, with Rocky Mountain elk taking an intermediate position.

I am unaware of archaeological or paleontological reports estimating how long Roosevelt elk have possibly inhabited north-coastal California or western Oregon. On the coastal plains and western slopes of the Cascade Mountains in Washington State, elk remains have been dated up to 5,000 to 7,000 years ago (Harpole & Lyman, 1999). Using this information, I do not consider elk a recent colonist of the RNSP. Staying on the conservative side, I suspect elk have inhabited north-coastal California for at least 1,000 years.

Written accounts mentioning elk in north-coastal California begin with the Jedediah Strong Smith exploration of the area in 1828 (Dale, 1918; Sullivan, 1934). Smith and his assistant Harrison G. Rogers kept daily journals and were the first Europeans to visit the area. Reading both journals was fascinating. I could pick out, or thought I could pick out, some of the locations that the party traveled in northern California. What I find as a place of great beauty and joy was truly a challenging and dangerous landscape to European explorers in 1828. The large trees were noted by Rogers but he did not speak about them with reverence or awe as we do today. The coastal Redwood trees (called white cedar in Rogers's journal), both standing and fallen, might have been viewed as obstacles to get around. Also, I was struck by how Rogers's observations of elk were similar to mine during my first elk study on private timberland north of the RNSP. Small bands (groups) of elk were commonly encountered and elk sign (tracks, feces) was even more frequently encountered. I strongly suspect that elk were ubiquitous at that time.

The next account I read was written by Isaac Jones Wistar (1937) from his travels in California during the gold rush in 1849–1850. He noted the abundant herds of elk in what is now the Bald Hills area in the southeast part of the RNSP. Also, he talked about hunting grizzly bears (*Ursus arctos*), which also seemed to be common at the time. I note grizzly bears because they prey on elk, particularly on juveniles (Griffin et al., 2011).

The history of large predators of elk in the RNSP is intriguing. Grizzly bears were in the area until they were shot out by 1890 (Dasmann, 1964). Storer and Tevis (1955) report 1868 as the last scientific record of grizzly bears in north-coastal California. Black bear were also common during the Jedediah Smith exploration, persisted through the period of overexploitation in the 1800s and early 1900s, and are ubiquitous today (Dale, 1918; Dasmann, 1964; Early, 2010; Sullivan, 1934). Puma occupied north-coastal California for some time and were probably abundant before Europeans arrived. Around 1900 puma abundance probably changed. The State of California offered bounties for puma between 1907 and 1963 (Dasmann, 1964). If you treat the number of bounties as some indicator of abundance, puma abundance declined in the north-coastal counties from 1907 to 1950. There were 613 bounties between 1908 and 1910, 245 from 1928 to 1930, and 172 from 1948 to 1950 (Dasmann, 1964). After 1950 puma along north-coastal California became more apparent (Allen, 2014; Fifield, Rossi, & Boydston, 2015).

The gray wolf (*Canis lupus*) in north-coastal California is its own story. It is not certain whether wolves occupied the area that is now the RNSP before 1828. Part of the issue with historical documentation of wolves is that the writer might confuse wolves with coyotes (*Canis latrans*) or refer to coyotes as wolves (Schmidt, 1991). Confusing black and grizzly bears was not nearly as much of an issue simply because of the lethal danger to human life from the much more aggressive and unpredictable behavior of grizzly bears. Perhaps the only way to have some degree of confidence that the writer could accurately identify wolves is if their accounts mentioned the three forms of canids that could have been in the area at the time: the large wolf, the intermediate-sized coyote, and small foxes (Schmidt, 1991). Schmidt (1991) stated that wolves were reported in 1828 to be in what is now the RNSP. Schmidt's statement is based on the written account of May 22, 1828, in Jedediah Smith's journal (Sullivan, 1934). In describing the species sighted, he noted some large and small wolves and foxes. Dasmann (1964) made no reference to wolves in his report. Stephanie Shelton, a student who helped out on field work in 2004, compiled historical range maps of the gray wolf in North America (Fig. 1.2). Her findings confirmed the anxiety and apprehension about the historical distribution and abundance of wolves in north-coastal California (Shelton & Weckerly, 2007). Going back to 1828, I did not find evidence to suggest that gray wolves inhabited the RNSP.

Following the heavy exploitation of elk throughout California in the late 1800s and early 1900s (Ellsworth, 1930), the area that would become the RNSP was one of the few remaining areas in the state with Roosevelt elk or any subspecies of elk (McCullough, 1969). Because of the restricted range at that time, there were calls to conserve elk (Orr, 1937). Perhaps because elk were scarce at this time, written accounts of observations of elk in what became the RNSP started to emerge in 1925. Barnes (1925a) wrote a short paragraph delineating elk distribution in an area roughly 6.4 km by 32 km north of Orick, California, which is now in the RNSP. In the present day, the area of reference is roughly U.S. Highway 101 on the east, the Pacific Ocean on the west, and north from Orick to the Klamath River. Barnes (1925b) wrote another short paragraph reporting in the neighborhood of 100 elk in the area of reference. One sentence in that paragraph (Barnes, 1925b, 90) states, "The country where they range is very brushy and heavily timbered and it is very difficult to obtain reliable information as to the exact numbers." By this time, Prairie Creek Redwoods State Park included a meadow in the area of reference that was purchased from the Boyes family (Veirs, 1987). The Boyes Meadows, along with nearby timberland, was purchased by Save the Redwoods League in 1932 and gifted to the State of

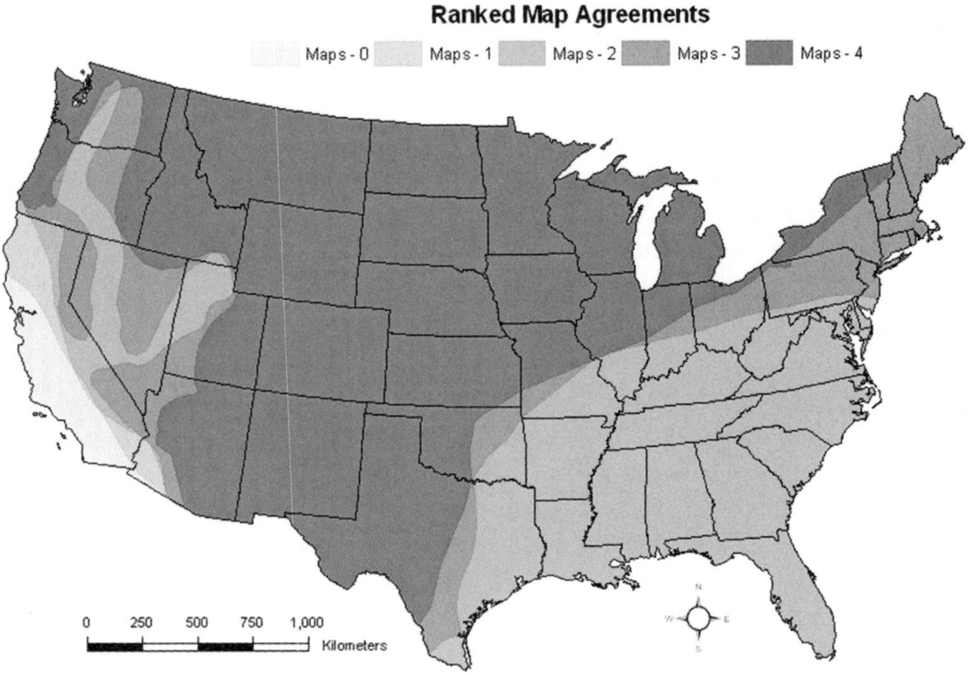

FIGURE 1.2. The ranked agreements among four historical range maps for the gray wolf (*Canis lupus*) in the conterminous United States (Shelton and Weckerly, 2007, Figure 2). Places where the maps agreed on the historical (1500 to the present) range among none, one, two, three, or all four maps. Note the extent of map disagreement in north-coastal California.

California. The next observations of elk in this area were by Orr (1937). Between October 17 and December 19, 1935, Orr was in the area of reference. His observation from his short time in the field and talking with local residents was that there were 200 elk. He suggested that some elk, in the Prairie Creek drainage, Boyes Meadows, and nearby ridges, leave in the autumn after leaf fall and move toward the coast, where some elk remain throughout the year. The veracity of these observations, regrettably, will never be known.

In an area centered on Boyes Meadows, Roosevelt elk went from scarce in the 1930s to abundant in the 1940s (Fig. 1.3) (Harn, 1958). A few elk were noticed in Boyes Meadows in 1937 (Dasmann, 1964; Harn, 1958). The elk might have shown up in Boyes Meadows after a fire in a nearby forest (Bentley, 1959). In the late 1930s and much of the 1940s, all accounts and available data suggest that the elk population grew rapidly. In November 1949 Raymond Dasmann,

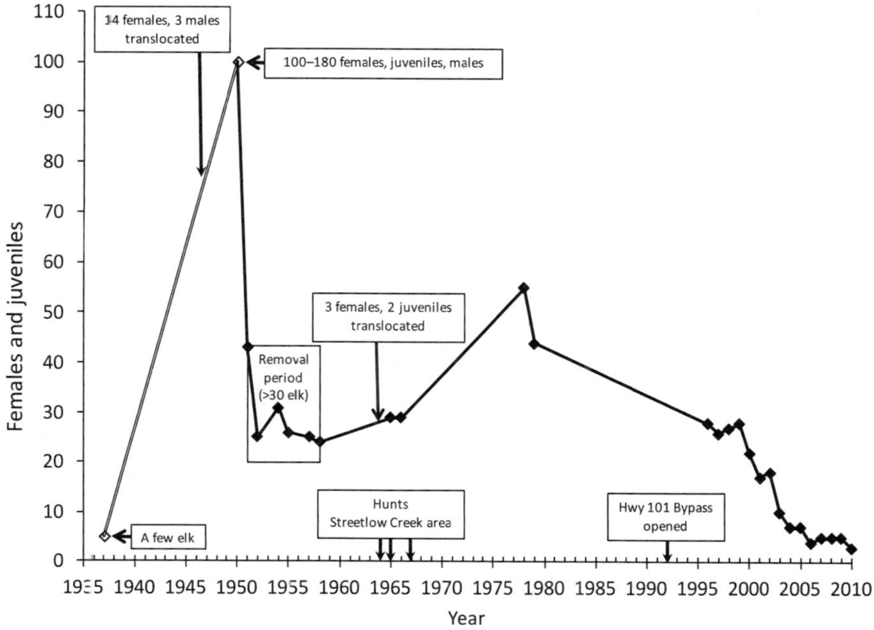

FIGURE 1.3. A time series of the number of females and juveniles in the Boyes herd between 1937 and 2010. The time series and associated information were put together from reports (Dasmann 1964; Mandel & Kitchen 1979), theses (Franklin 1968; Harn 1958; Lemos 1971; Logsdon 1965), and population surveys I conducted from 1997 to 2010.

then a faculty member at Humboldt State College, counted more than 100 elk in Boyes Meadows in one day (Harn, 1958). Accounts suggest that at peak abundance there were a minimum of 100 and a maximum of 180 elk in the Boyes Meadows population. To be less dramatic, I report 100 as peak abundance (Fig. 1.3). Elk deaths due to malnutrition were particularly evident in the winter of 1949–1950 (Harn, 1958). After the population peak in 1949, there was a crash from 1950 to 1951. In 1952 there were 25 females and juveniles counted in Boyes Meadows. After the population crash there was a period between 1950 and 1957 when management activities were directed at reducing population size and alleviating elk grazing pressure on the vegetation in Boyes Meadows (Bentley, 1959; Harn, 1958, Harper et al., 1967). One means of reducing population size was to live-trap elk in Boyes Meadows and translocate them to other parts of California where Roosevelt elk had been extirpated. Elk were also dispatched if they were emaciated and near death. Between 1957 and 1967, elk removal activities were continued because there were additional translocations

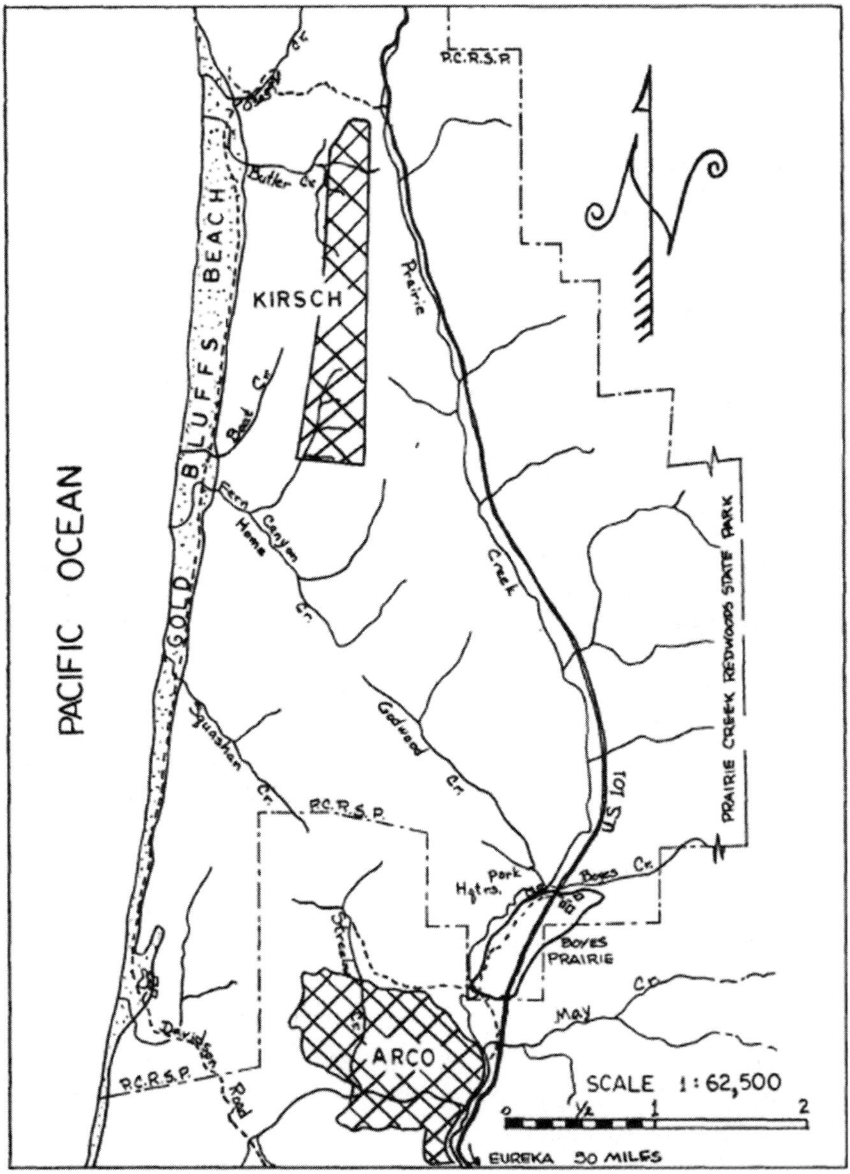

FIGURE 1.4. A map from the thesis of Lemos (1971, Figure 2) showing the privately owned (at the time) Arcata Redwood Company (ARCO) land that contained the bulk of the Streetlow Creek area that had a public hunt in 1964, 1965, and 1967. Timber harvest occurred in 1958 and 1960. The crosshatched area labeled Kirsch had timber harvested in the early 1950s.

and three public hunts that occurred on what were then private lands directly south of Boyes Meadows in the Streetlow Creek area (Fig. 1.4). From 1967 to 2016 there were no more translocations of elk inhabiting Boyes Meadows. The data points from 1997 to 2010 in Figure 1.4 were from my population surveys and will be analyzed in more detail in Chapter 5. I also note on Figure 1.3 the opening of the Highway 101 Bypass in 1992. The Bypass created new elk forage habitat along the shoulder of the highway that was used by the Boyes population during the time of my study. With the opening of the Bypass, Highway 101 no longer went through Boyes Meadows as it had done since its construction in the 1920s. Highway 101 bypassed Boyes Prairie to the east (Fig. 1.5).

Boyes Meadows was not the only area of the RNSP that had and has elk (Julian et al., 2013; Mandel & Kitchen, 1979; Starns et al., 2015). Timber harvesting on private lands in the Prairie Creek drainage and elsewhere on lands that would become the RNSP increased in activity between 1950 and 1968 and continued until the mid-1970s. In 1968 the U.S. Congress created Redwood National Park and expanded land holdings in 1978. Redwood National Park and the state parks adjacent to, or more or less surrounded by, the Redwood National Park began joint management in 1994. Before and shortly after the creation of the Redwood National Park, elk distribution expanded and abundance increased in the southern part of the RNSP because of the increase in forage habitat brought about by timber harvesting (Lemos, 1971; Mandel & Kitchen, 1979).

Some private land still existed in the Prairie Creek drainage when Redwood National Park was created. In 1991 the National Park Service acquired a part of that private land, the Davison Meadows complex (Fig. 1.5, 1.6). Again, Save the Redwoods League purchased the property and deeded the land to the National Park Service. The Davison Meadows complex originated from forest land that was cleared beginning in 1890. After the trees were felled, pasture for cattle was created. Cattle grazed these meadows for about the next 100 years (Starns et al., 2015). Elk quickly became apparent in the Davison Meadows complex shortly after the land became National Park Service property in 1991. The story of the Davison elk is told in Chapter 4.

From 1997 to 2014 there were five elk herds (populations) in the Prairie Creek drainage (containing Boyes Meadows and the Davison Meadows complex), along the coast in the Gold Bluffs area west of Boyes Meadows and in the lower Redwood Creek drainage by the town of Orick (Fig. 1.5). I collectively refer to these herds as the lower Redwood Creek and Prairie Creek herds for the sake of convenience. Because elk had become ubiquitous near Orick

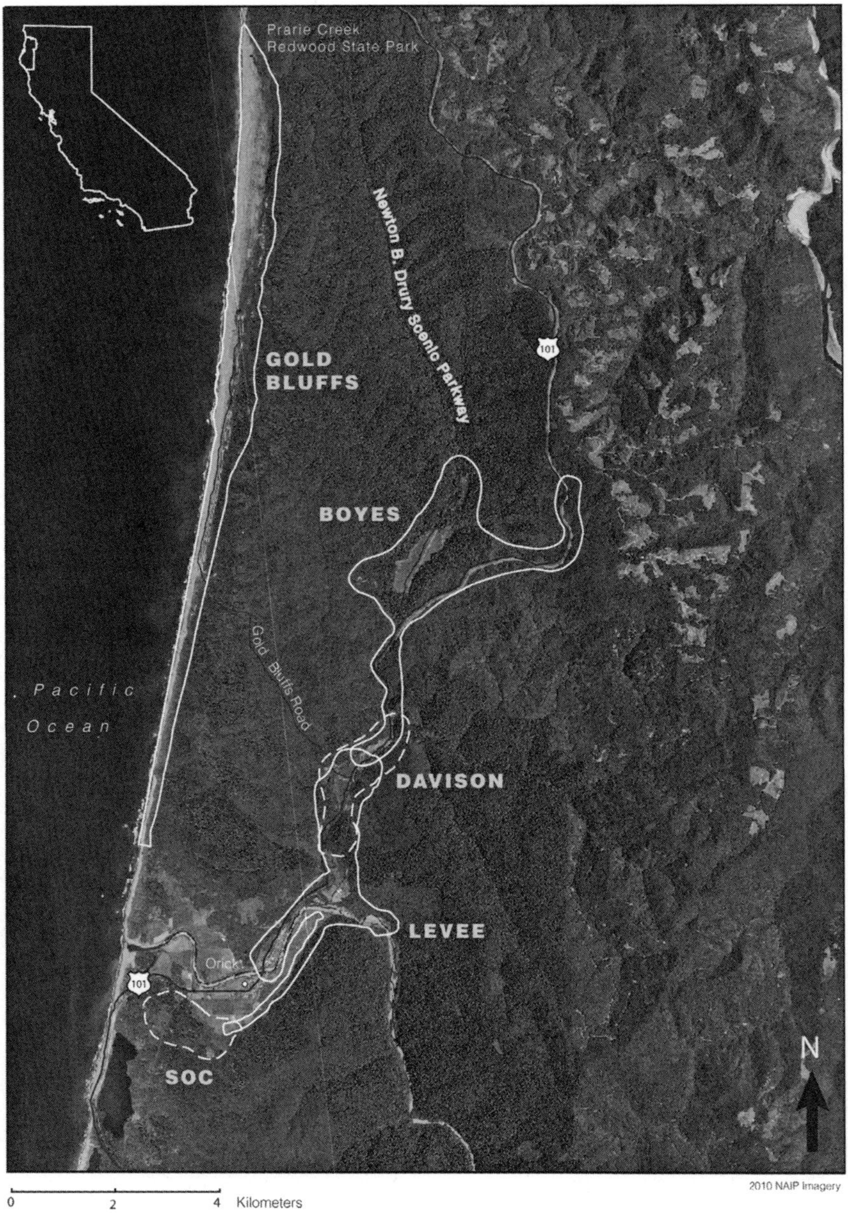

FIGURE 1.5. Aerial photograph showing the Prairie Creek drainage and the areas occupied by elk herds in the southwestern part of Redwood National and State Parks. The Bypass is Highway 101 east and north of its junction with the Newton P. Drury Scenic Parkway. The polygons delineate the boundaries of the five elk herds inhabiting this area of the RNSP from 1997 to 2014 (Julian et al., 2013; Starns et al., 2015).

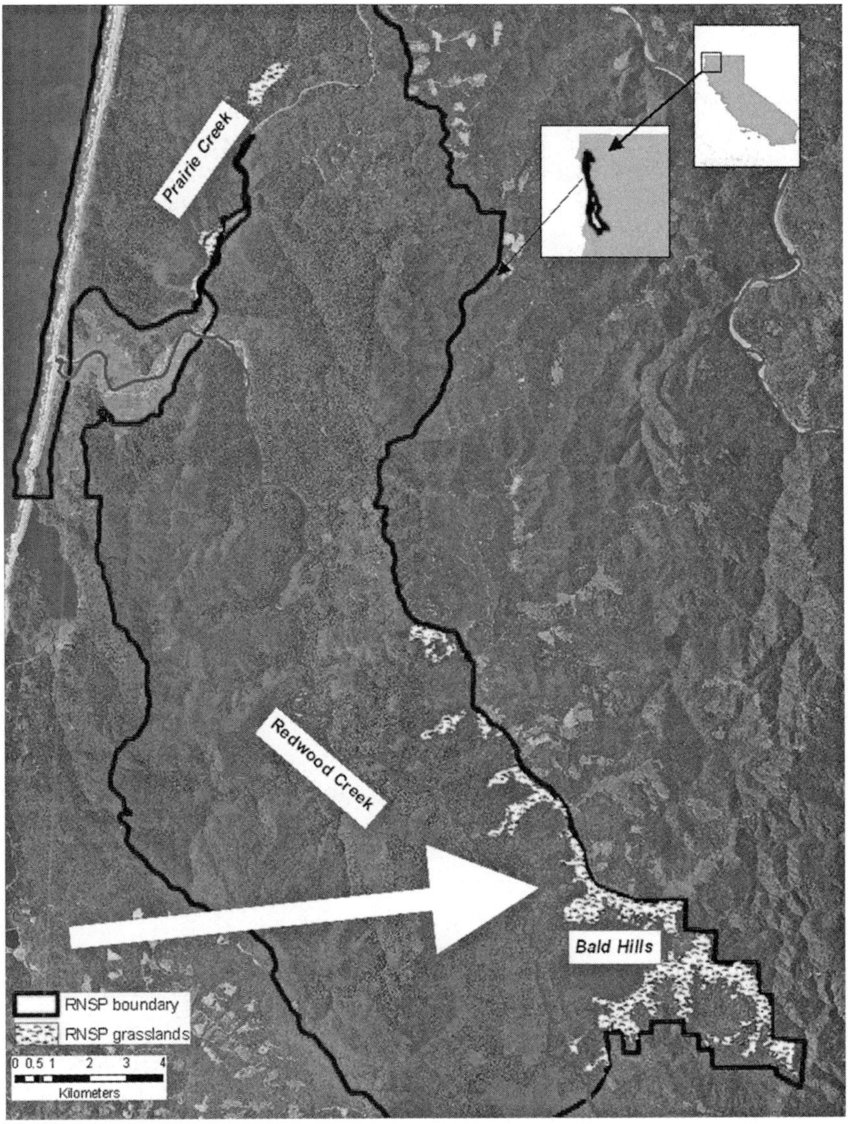

FIGURE 1.6. Map of the southern part of Redwood National and State Parks showing the Prairie Creek drainage, with its two meadows (Boyes to the north, Davison complex to the south), lower Redwood Creek area, and the Bald Hills Meadows in the southeast portion of the RNSP. The black lines represent RNSP boundary, and the white stippled area represents meadows. The white arrow is a route that elk might have used to recolonize the Bald Hills from the Big Lagoon area. Mark Ricca made the map.

since at least the 1970s, the RNSP built a fence in 1990 as a barrier to prevent elk from foraging in pastures on private lands that were south and east of Orick (Hofstra, 1997). About 10–20 km to the southeast of the Prairie Creek drainage are large meadows inhabited by the Bald Hills population (Fig. 1.6). Since the cessation of logging in the late 1960s and 1970s on the lands that became the RNSP, plant succession in these cutover areas has resulted in closed-canopy, second-growth forests that contain little grass and other elk forage (Cook, Cook, Davis, & Irwin, 2016; Harper et al., 1967; Weckerly, 2005). Much of the areas inhabited by the present-day elk populations contain meadows. Meadows, by the way, contain the bulk of elk forage in the contemporary landscape of the southern part of the RNSP (Weckerly, 2005).

The Bald Hills now contain the most abundant population in the RNSP (Starns et al., 2015). Piecing together the particular history of elk in this part of the RNSP has not been attempted. As stated earlier, elk were ubiquitous and perhaps abundant in the large meadows of the Bald Hills before Europeans arrived (Hofstra, 1997). Around 1850 homesteading on the Bald Hills began and the meadows were grazed by cattle, horses, and mules (Hektner, Martin, & Davenport, 1983). In addition to being a food source, elk were probably viewed as competitors for grass and, I would guess, mostly ended up in a cooking pot. Sheep evidently became the predominant domestic grazer in the late 1800s until the 1940s (Hektner et al., 1983). To make sheep ranching economically viable, natural predators like black bear, bobcat (*Lynx rufus*), coyote, grizzly bear, and puma must be kept at bay. Kept at bay usually means predator control. Sheep are herd animals that are both docile and placid. Intelligent and fierce are not apt descriptors of sheep, albeit an anthropomorphic choice of words on my part. Black bear and puma populations were probably not abundant between the late 1800s and the 1940s in Bald Hills. Grizzly bears were certainly extirpated (Dasmann, 1964). In the 1950s most of the Bald Hills grazed by sheep were replaced by cattle, which continued to graze the Bald Hills until being phased out when the area was purchased by the National Park Service. The last special use permit that allowed domestic livestock to graze the Bald Hills ended in 1981 (Hektner et al., 1983).

Also, to my knowledge, no one has explained how elk recolonized the Bald Hills. I offer three possibilities. Residual elk is one possibility. Elk might not have been completely extirpated from the Bald Hills. A few elk might have persisted in the Bald Hills when domestic livestock were present. This is a possibility that has to be considered given the lack of documentation in the Bald Hills between 1850 and the mid-1900s. Another possibility is nomadic elk. In

Humboldt County, large meadow complexes are prevalent about 15–30 km inland from the coast from Bald Hills, extending to the south for many kilometers. In such a large area, a few elk might have persisted. Barnes (1925a) noted six elk reported to him by a farmer near Iaqua (spelled Iquay by Barnes), which is now a ghost town. Iaqua was approximately 50 km to the south of the Bald Hills. Harn (1958) stated that elk had recently been spotted in the Bald Hills. There might have been a few elk that persisted through wandering over a large area that included the Bald Hills. Site fidelity is the standard in both migratory and nonmigratory elk. But elk can display nomadic movement behavior and, perhaps, the environmental setting at the time was not conducive to site fidelity (Ackerman, Kuck, Merrill, & Hemker, 1984; Bowler & Benton, 2005; Boyce, 1991; van Beest, Vander Wal, Stronen, Paquet, & Brook, 2013). Group sizes were probably small and elk in small groups might have been more apt to perceive risks from predators and people. Domestic livestock grazing might have dominated the landscape so that food supplies were not rich, or predictable, and displacement by cattle might have also occurred (Weckerly, 2005). Yet, once domestic livestock were removed, risks to elk became less and individuals established fidelity to the Bald Hills and became abundant. Recolonization from Big Lagoon elk is the third possibility. In the 1960s and 1970s much of the south side of Redwood Creek drainage, opposite the Bald Hills, was logged of its old-growth timber. The logging, of course, created elk forage habitat for the next 10–20 years after the trees were harvested. Directly to the southwest of the Redwood Creek drainage is the Big Lagoon area where elk were ubiquitous at the time (Harper et al., 1967; Hofstra, 1997; Mandel & Kitchen, 1979). Elk possibly dispersed north from the Big Lagoon area to the south slope of Redwood drainage, eventually crossed Redwood Creek, and recolonized the Bald Hills about the time that grazing by domestic livestock ceased (Fig. 1.6). However elk became ubiquitous again in the Bald Hills: two adult females were reported in the Bald Hills on October 30, 1978, and another three adult females and one subadult male were sighted on December 27, 1978 (Mandel & Kitchen, 1979). I used the December 27 sighting as the source of the population that now inhabits the Bald Hills.

In summary, along north-coastal California it seems that elk were once common, then became less common, and today are common again. In Chapter 2 I present the details of how I actually measured distribution and abundance during my 20-year study. It is also convenient to talk about other widely used techniques and my analyses about some of the nuances involved in the kind of abundance estimates I report throughout the book.

2

Techniques, Counts, and Population Estimates

A day filled with bad elk stalks
is still better than 100 days of just about anything else!

— Butch Weckerly, field notes, January 5, 2007

Some data were collected year-round (e.g., Fig. 0.5) but the bulk were collected in the winter, primarily January but also February through April. Winter was the field season for two main reasons. One, in winter spatial patterns should be more influenced by resource use and less by reproductive activities. In late spring and early summer, females with neonatal offspring sometimes use habitat that offers sanctuary from predators at the expense of using habitat that provides forage (Bowyer, 2004). In autumn, males and also females engage in reproductive activities that override space use to acquire resources (Bowyer, 1981). Also, in early winter green feed is available but not abundant (Fig. 0.5). What nutritious food is available is mostly in meadows and not forests because leaves of most palatable browse (shrubs and trees) species abscised the previous autumn. If partitioning of meadows is going to occur, it should be apparent in winter when forage is limited (Kolbe & Weckerly, 2015). The other reason had to do with locating juveniles during population surveys and classifying elk into age–sex categories. The categories were juveniles (0–1 years old), subadult males (1–2), adult males (> 2), and adult females (> 1). In summer, herbaceous vegetation in meadows is tall but juveniles are small. Locating juveniles in summer and early autumn can be challenging. In January juveniles are much larger than in summer and much of the meadow vegetation is short. Furthermore, juveniles have not grown to the extent that it is difficult to distinguish them from subadult and adult females. I could not readily distinguish subadult and adult females over the distances I made observations. The most

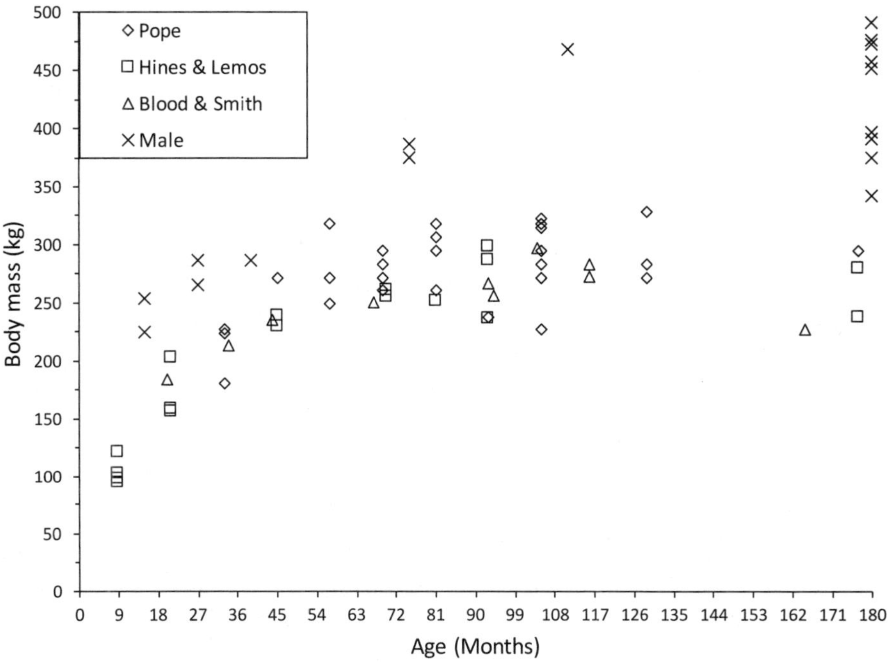

FIGURE 2.1. Age and whole body mass of female and male Roosevelt elk. Data were obtained from studies (Blood & Smith 1984; Hines & Lemos 1979; Pope 1994) reporting age and female body mass. The male body mass data were collected by Mark Ricca from males harvested during the Smith River hunt, Del Norte County, California, 1993–1996. Male age was determined by counting cementum annuli rings from an incisor extracted from the harvested elk (Asmus & Weckerly, 2011). The male data points reported as 180 months in the graph were actually animals that had *no age data*.

likely distinguishing feature was size but the size differences can be slight in January (Fig. 2.1).

Readily observable differences in size are needed to reduce errors in classifying females into these two age categories. Males, in comparison, can be distinguished to subadult or adult age categories based on gross antler morphology. Subadult males were assumed to have short, unbranched antlers whereas adult males were assumed to have longer antlers that branched or had tines. Furthermore, in January both subadult and adult males have fully developed, hard antlers. Larger, older adult males tend to cast antlers first, often in February, and subadult males might not cast antlers until May. Finally, there was the reality of the shoestring budget. I moved to Texas in late summer 2002 to take on a new job. From 2003 to the present I usually had funds to travel between Texas and California once a year.

Techniques that I used more often or were a large part of a chapter are presented here. If I used the technique once, I will describe it when presenting the analysis and findings.

DISTRIBUTION

Location of solitary elk, groups, and herds were recorded in Universal Transverse Mercator (UTM) coordinates (White & Garrott, 1990). For the first 5 years of the study locations were determined from U.S. Geological Survey topographic maps (1:24000 scale). Since 2003 I had the luxury of using global position system (GPS) units to obtain elk locations. Either I went to the spot where the elk were found or I calculated the elk location from my UTM coordinates, a compass bearing to the elk and measuring the distance between me and the elk with a rangefinder.

Forage abundance in meadows was measured both indirectly and directly. Indirectly, forage abundance was assessed using the Normalized Difference Vegetation Index (NDVI). The NDVI is measured from satellites and is the ratio of near infrared to red light reflected by vegetation. Plant greenness is what NDVI directly measures yet NDVI correlates well with plant biomass and plant growth (Hebblewhite et al., 2008; Lendrum, Anderson, Monteith, Jenks, & Bowyer, 2014; Muñoz et al., 2010). Details of collating NDVI values are available in Starns et al. (2015) but summarized here. NDVI values were measured in 30-m^2 pixels in meadows from 1989 to 2011. Satellite images were taken 16 days apart, but cloud cover on days that satellites passed over the RNSP was an issue. Consequently, some months did not have NDVI values because images were not useable or there were months that had only one usable image. When my graduate students were fortunate and had two useable images in a month, they averaged the NDVI values for the month.

The food supply in January was directly measured in January 2005–2016, in 570 quarter-m^2 plots (550 such plots in 2005). A quarter-square-meter area was roughly the area that an elk took bites from before moving its head to feed elsewhere. Ten to forty plots were arranged along randomly selected transects. Plots were spaced 10 m apart along transects, and transects were at least 20 m apart. Transects and number of plots along transects within Bypass, Boyes, and Davison Meadows were selected in proportion to the total area of meadows. In the larger Boyes and Davison Meadows, sectors were delineated to distribute transects. Establishment of sectors considered the need to survey all parts of meadows and to capture environmental heterogeneity that might be perceived by elk from accessibility to people and vegetation biomass. Figure 2.2 shows the sectors in Davison Meadows.

In every plot, I measured canopy cover and height of grasses, forbs, and shrubs palatable to elk (Peterson, 2016). I visually estimated canopy cover of grasses, forbs (herbaceous dicotyledonous plants, e.g., weeds) and shrubs using Daubenmire (1968) coverage classes. I also measured vertical height of grass in eight equidistant spots in each plot. Vegetation height was measured to the nearest centimeter with a ruler.

Calibration regressions estimated the biomass of grass as well as forbs and shrubs in each of the quarter-square-meter plots (Peterson, 2016). Dried biomass of the plant types was directly measured in a different set of 130 quarter-m^2 plots randomly placed throughout Davison, Boyes, and Bypass Meadows. After measuring canopy cover and vegetation height in each of these plots, the vegetation was clipped to ground level, sorted into palatable grasses, forbs, and shrubs, dried for 30 hrs at 60°C, and then weighed to the nearest hundredth of a gram.

I analyzed general linear models or calibration regressions to estimate biomass. Separate regressions for grasses and forbs and shrubs were estimated because of different shape geometries and because forbs and shrubs were scarce (mean forb and shrub:grass biomass ratio was 0.06, sample size (n) = 130). The regression for grass biomass had predictors of mean height, mean height squared, and grass cover. The regression for forb and shrub biomass had forb and shrub cover as the predictor. The regressions did a reasonable job of predicting forb and shrub ($r^2 = 0.33$, $n = 96$) and grass ($r^2 = 0.84$, $n = 130$) biomasses. Sample size of forbs and shrubs was less than grass because there were 34 plots with no forbs or shrubs. Predicted values from each regression were summed to estimate forage biomass in each of the quarter-square-meter plots (Peterson, 2016).

ELK BEHAVIOR

Social bonding or low fission–fusion dynamics among females, juveniles, and subadult males is a hallmark that essentially defines herds of Roosevelt elk (Franklin & Lieb, 1979; Weckerly, 1999). Female young often remain with their natal herd their entire lives (Franklin & Lieb, 1979). Males usually remain in the natal herd as juveniles and as subadult males. Males disperse from their natal group between 2 and 3 years of age. Males might leave their natal herd for a variety of reasons. For instance, their foraging needs diverge from females

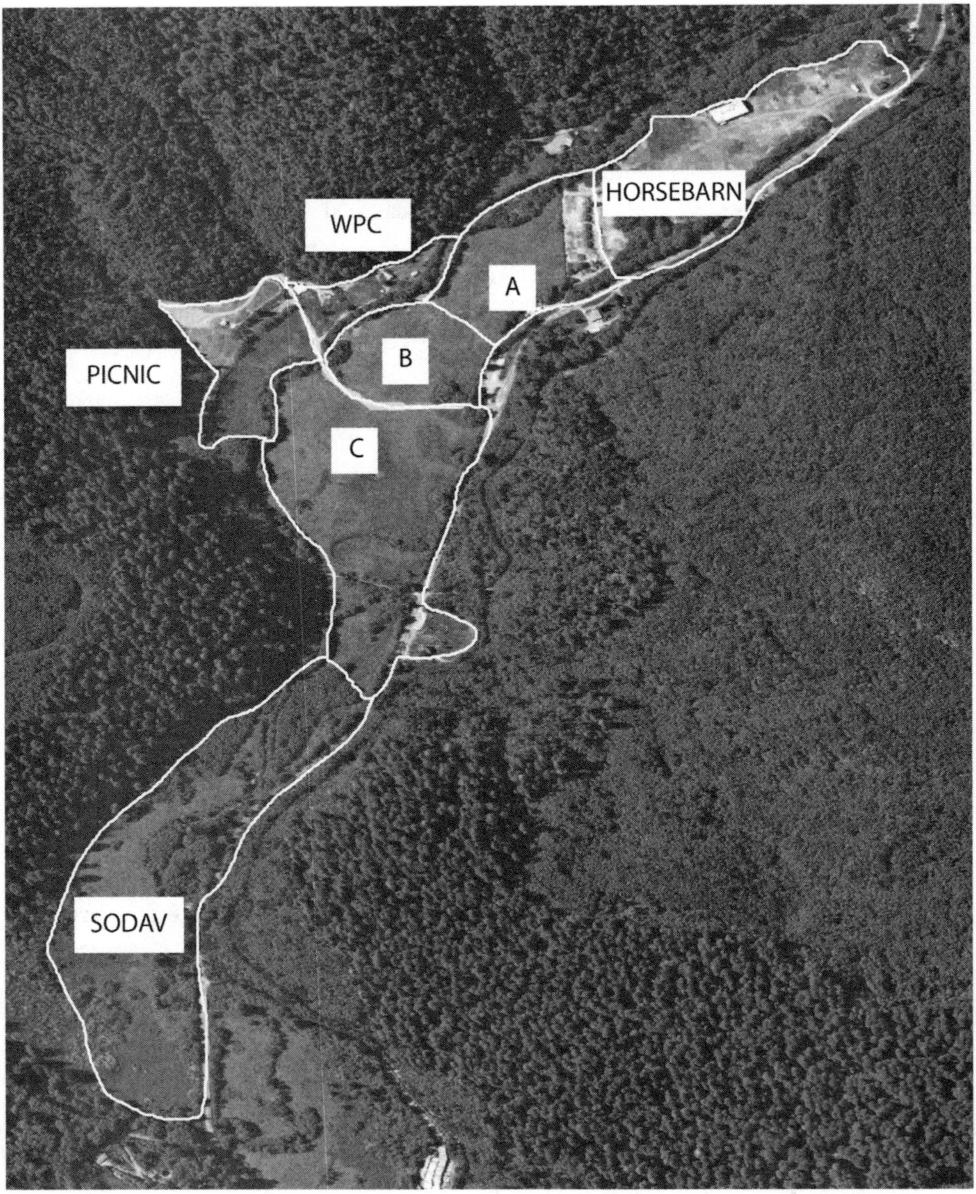

FIGURE 2.2. Aerial photograph of Davison Meadows showing the sectors in each meadow where vegetation was measured.

because they are larger and still growing (Fig. 2.1), to reduce the possibility of breeding with female kin, or to reduce harassment, which can be violent, from sexually mature males during the mating season (Loe, Mysterud, Veiberg, & Langvatn, 2009; Long, Diefenbach, Rosenberry, & Wallingford, 2010).

The first work revealing the extent of social bonding among females was a series of studies conducted between May 1964 and May 1968 (Franklin, 1968; Lieb, 1973; Logsdon, 1965). These studies were master's theses conducted on the Boyes and Gold Bluffs herds. In these two herds, virtually every elk was ear-tagged in such a manner that tagged elk could be uniquely identified. Logging occurred at the time and 0–10-year-old forests were available to the south and east of Boyes Meadows (Franklin & Lieb, 1979). Herds in addition to the Boyes and Gold Bluffs herds used these young forests but they were infrequent visitors to Boyes Meadows and not considered part of the Boyes herd. There was a total of 93 juvenile, subadult, and adult females that were ear-tagged in the Boyes and Gold Bluffs herds, of which 12 went missing. Their fate as of May 1968 could not be confirmed as alive, dead, or trapped and translocated. Among the remaining 81 tagged animals there was only one instance where a 2-year-old female dispersed from the Boyes herd. This elk moved to a herd that inhabited the Streetlow Creek area to the south of Boyes Meadows and then remained with that herd.

The theses of Logsdon, Franklin, and Lieb also showed that most females, juveniles, and subadult males aggregate in the same group between late autumn, after the mating season, to shortly before the season of parturition. During the mating season, the activities of males can result in herds being split into smaller groups. Just before giving birth, females separate themselves from herds. After parturition mother and offspring remain by themselves for up to four weeks (Kristan, 1992). Seclusion from the herd probably evolved to facilitate mother–offspring imprinting and to reduce predation on neonates (Bowyer, 2004; McCullough, 1979).

I conducted a study in 1996 and 1997 reaffirming the strong social bonding of females, juveniles, and subadult males in late autumn through early spring (Weckerly, 1999). I studied the Boyes and Davison herds. At the time, there was only one ear-tagged elk. I estimated strong social bonding through analysis of group sizes, the presence of a few naturally marked elk, and the one ear-tagged elk and counts of the total number of elk in the area. A group was defined as elk displaying coordinated activity within 50 m of one another (Weckerly, 1999).

I used focal animal sampling to measure individual elk behavior (Weckerly, 1993). Behaviors included time the head was in the feeding position (HFP), number of bites taken (cropping), and number of chews (Weckerly et al., 2001). Most data were collected when animals were in groups, although solitary elk were measured when encountered. My selected focal animal was observed and measured between 4 to 20 mins, depending on the particular project goal and behavior that was being measured. In addition to determining its UTM coordinates, standardized descriptors of meadows and sectors were recorded, as was group size. Depending on the behavior measured, the time other elk in a group were within one body length of the focal animal was also measured. I selected focal animals in groups that had an unobstructed view. When I took multiple focal observations from the same group in a day, care was taken to reduce measuring the same animal repeatedly. Elk in different parts of the group were selected; however, if elk could be uniquely identified by morphological features, then the same animal was not measured twice in the same day. If the focal observation had to be stopped before the minimum time limit, the observation was discarded. I collected focal animal data from 2005 to 2014.

I also measured group behavior of the Davison herd. I followed this herd from dawn to dusk in January 2009–2012. At this time of year, the short day lengths allowed me and my graduate students to collect data continuously for up to 9–10 hrs. I collected data in 8 days in 2009, 9 days in 2010, 14 days in 2011, and 11 days in 2012. If the group went into a forest, which made it difficult to observe all individuals, for longer than 2 hrs, then data collection was terminated and the data discarded for that day. Seven or fewer hours of observing and recording activity data in a day was viewed as possibly resulting in biased measurements during daylight hours. At five-minute intervals, group location (UTM and sector), size, and activity state of each individual was recorded. Activities were foraging, loafing (standing up and displaying no other obvious behaviors), being bedded, and other.

A group by definition excludes solitary animals. Yet, it is possible that elk can be solitary or leave groups and become solitary for a time. Males in particular did this (Weckerly, 2001). To summarize group size, I often calculated typical group size (Heard, 1992; Jarman, 1974). Typical group size attempts to capture the gregariousness of a typical animal. An appealing feature to typical group size is that the measure is more resistant than mean group size to the influence of a few solitary animals or a few large groups.

STATISTICAL ANALYSES

The primary statistical tools used to analyze data were mixed effects models and linear models such as analysis of variance, correlation, and regression (Pinheiro & Bates, 2000; Sokal & Rohlf, 2012). Categorical predictors had the number of categories minus one possible dummy variable. Dummy variables were coded 0 (reference category) or 1. When there were more than two categories, the category coded 0 for all dummy variables was the reference category. Linear and linearized mixed effects models were particularly well suited when the response variable was repeatedly measured from the same group or in the same area, or when a predictor was a random factor. By definition, a random factor is a subset of categories drawn from the realm of all possible categories for the factor (Sokal & Rohlf, 2012). The proportion of variation in the response variable that can be accounted for by the fixed effects (marginal r^2) and the fixed and random effects (conditional r^2) also were calculated for mixed effects models (Nakagawa & Schielzeth, 2013). Unless stated otherwise, intercepts random effects were estimated for each random factor in a mixed effects model. In program R, package lme4 was used to estimate mixed effects models using restricted maximum likelihood estimation. Degrees of freedom were calculated using Sadderthwaite's approximations in the lmerTest package (R Core Team, 2014).

I used Akaike's Information Criterion (AIC),

$$-2^{*}\text{log likelihood} + 2^{*}\text{number of parameters estimated} \qquad 2.1$$

to select models when multiple models were estimated from a data set (Burnham & Anderson, 2002). I used AIC model selection because it was versatile. When models have the same response variable and predictors are fixed, the AIC approach is applicable. For brevity, I used the initials LL for log likelihood and nPar as the label for number of parameters. Because my data sets were small (n/nPar < 40), I calculated the AIC corrected for small sample size (AIC_C)

$$-2^{*}\text{LL} + 2^{*}\text{nPar}[n/(n - \text{nPar} - 1)]. \qquad 2.2$$

The smaller the AIC_C the more likely the model summarizes data, given the number of parameters estimated by the model. Models with more parameter estimates are penalized to have higher AIC_C's. Parsimony, therefore, is also a criterion in model selection. One issue in some of my model selection analyses was when nested models differed by one parameter as well as differing by

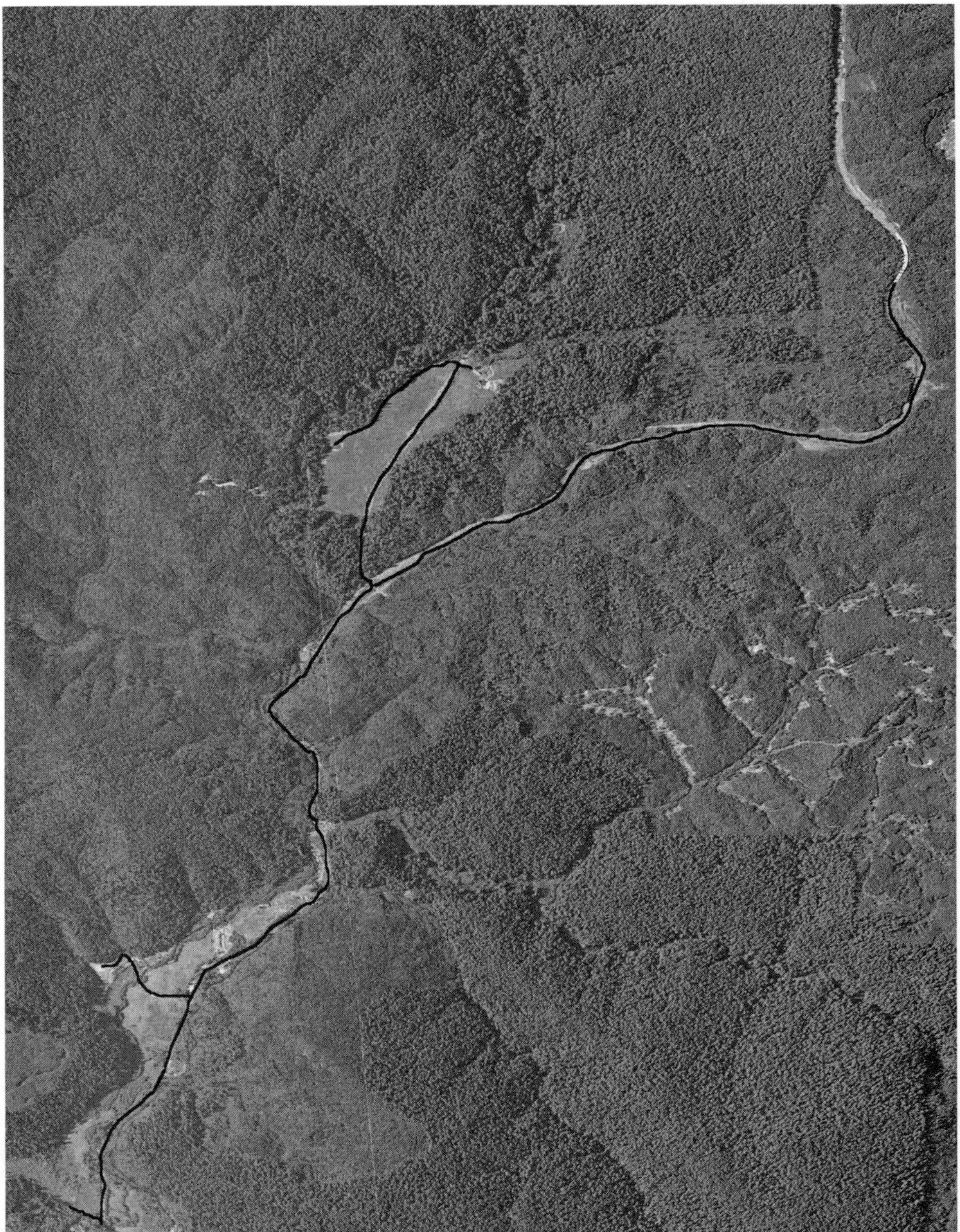

FIGURE 2.3. Aerial photograph of the lower Prairie Creek drainage showing the route driven on each population survey.

< 2 AIC_C units ($\Delta < 2$). Nested models are either complete or reduced. Reduced models contain a subset of the same parameters estimated in the complete model. The model or models with one additional parameter might be considered competing but not summarize the data any better than the model with one fewer parameter. This might be a simple consequence of the constant 2 in Formula 2.1 (Arnold, 2010). To further evaluate models with $\Delta < 2$ I also assessed LL (log-likelihoods) and r^2 values. The model with the smallest AIC_C might actually summarize the data somewhat better if the log likelihood is also somewhat larger than models with $\Delta < 2$ (Burnham & Anderson, 2002).

Using AIC_C model selection for mixed effects models is trickier because of random effects (Wolcott, Reitz, & Weckerly, 2015). In an AIC_C model selection analysis, models varied in fixed predictors but every model had the same random effects. Also, AIC_C statistics were calculated from models estimated using maximum likelihood. I then used restricted maximum likelihood estimators to report estimates and random effects from the selected model.

I also used statistical tools that involved bootstrapping and randomization of data. Bootstrapping is essentially sampling with replacement. In any bootstrapped sample, each observation has the opportunity to be randomly drawn multiple times from the original data set or distribution. Randomization tests require randomly rearranging data or, in other words, sampling without replacement. Both bootstrapping and randomization require a large number of iterations, 1,000 or 10,000, to summarize central tendency and variability or to obtain a null distribution. A null distribution is an attempt to remove any patterns in data so that it is possible to estimate the probability of a type I error and determine *P* values or confidence intervals (CI) (Sokal & Rohlf, 2012).

ABUNDANCE

I conducted population surveys each January and February in 1997 in the Prairie Creek drainage and Bald Hills. Surveys were conducted with a vehicle along a predetermined route in both areas (Starns et al., 2015; Weckerly, 1996). In the Prairie Creek drainage, the route was designed to allow me to view all parts of meadows in the lower part of the drainage (Fig. 2.3).

Beginning at dawn and for the next 1.75 hrs I drove the route and all elk encountered were classified to age–sex categories, and locations were recorded. When elk were detected, I often exited the vehicle to obtain a clear view of the elk. There were essentially two group types: female and male groups. Male groups were comprised entirely of adult males, and female groups had females, juveniles, and subadult males with or without a few adult males (Weckerly et al., 2001). Adult males entered and left female groups, not vice versa; when males did enter female groups, it was usually six or fewer males that did so. Group size and any naturally marked elk in the group were also noted. On days just before surveys began, the Prairie Creek drainage was searched for elk with morphological features that allowed me to uniquely identify them (Weckerly, 2007). Naturally marked elk were treated as marked elk so that I could estimate sighting probabilities and population abundances. Since elk were habituated to human activity in Prairie Creek drainage I could carefully scan elk at close distances (10–200 m) for prolonged periods of time to identify animals with unique features as well as classify elk and tally group size (Fig. 0.1a, 0.1b). If I thought I recorded data from the same elk multiple times in the same survey, only the first data recording was used. Except in 2 years, 10 surveys were conducted each year. In the 2 years of exception, 1998–1999, five surveys were conducted in each of those years (Starns et al., 2015).

In the Bald Hills it was not possible to survey all parts of all meadows because of the rugged terrain and the number of meadows, which often were large. In total, there was about 1,000 ha of meadow in the Bald Hills. I set up a standardized survey route that allowed viewing of as much of the Bald Hills Meadows as could possibly be done in 6–8 hrs (Starns et al., 2015). There were parts of the route that were driven in a vehicle and other parts that required hiking. Most of the time when elk were spotted from the vehicle I had to exit the vehicle to obtain a count. In 1997 and from 2002 through 2015, 6–10 surveys were conducted each year. In 2016 11 surveys were conducted. The goal was to conduct 10 surveys in each year but weather in some years prevented me from conducting surveys on 10 days in January. Rainstorms could result in a low cloud ceiling and reduced visibility; sometimes snow lasted for a week or so and was deep enough that elk stayed in forests at lower elevations below meadows. Eleven surveys were conducted in 2016 because the counts were noticeably lower than in 2015 (see the next section on estimates or counts of abundance). Finally, to increase the Bald Hills time series, I used counts reported by Mandel & Kitchen (1979) for 1978, by Grenier et al. (1990) for 1987, and from four surveys I conducted in 1996.

My initial observations indicated that solitary males or groups comprised entirely of males were rarely encountered in the Bald Hills. This observation concurs with the finding of Grenier (1991) that radio-collared males selected forest more often than radio-collared females in Bald Hills. When elk were in forests it was very difficult to find them. In forests, visibility was restricted, which meant I had to get close to elk to observe them. Usually elk heard or smelled me and fled before I could count them. My goal, therefore, was to count the female segment of the population in the open meadows in the Bald Hills. The female segment was comprised of juveniles, subadult males, and females. Those age–sex classes of elk associated with one another an overwhelming amount of the time (Weckerly et al., 2001).

The terrain and behavior of the elk in Bald Hills made the collection of count data more challenging than in the Prairie Creek drainage. Elk in Bald Hills were not as habituated to people as were the Prairie Creek elk. Typically, females in the Bald Hills walked or ran away shortly after they detected me. When a group was alarmed, females often clumped together, making an accurate count more challenging because elk obstructed my view of other elk. Group sizes of females could also be large (50–250). Large groups pose, perhaps, the greatest challenge to counting. Elk are usually spread over a large area; there are some to many elk in the group that are moving so it can be difficult to keep track of each individual when counting, and across a larger area there are more opportunities for elk to be obstructed by gullies, knolls, and vegetation. This means that when I counted a large group I might not detect every individual in the group or I might tally some of the elk more than once, particularly when the group was moving. My intent was to spot elk first before they spotted me so that I could plan a stalk to obtain an unobstructed view of bedded or foraging elk to count. Even when I spotted elk before they spotted me, the best view might be 1–2 km across a canyon. Sometimes a group was in a place where it was not possible to obtain an unobstructed view. In that situation, I might attempt to nudge the group into an opening where an unobstructed view was possible. Nudging involved approaching a group slowly. The elk would then detect the surveyor at a distance that would not alarm the elk and cause them to flee. The elk would then walk in another direction away from the person and into view of another surveyor that the elk had not detected. Sometimes nudging a group worked. The elk walked into an opening with an unobstructed view, a count was made, and I felt on top of the world. Sometimes nudging elk did not work and I was reminded once again of my limitations.

Taking digital photographs of large groups and counting the elk from the photograph was helpful. Unfortunately, the group had to be in a place where I could get close enough to take a photograph that had an unobstructed view of each and every elk in the group. In years when there were large groups, that might happen once.

FIGURE 2.4. A closeup view of a collar around the neck of a sedated female elk in the Bald Hills, January 2003. The collar was actually comprised of two collars: one collar was the very high frequency transmitter and the other collar was the GPS transmitter.

In the Bald Hills, the difficulty in counting elk and spotting marked elk, or elk with natural marks, so that I could estimate abundance like I did in the Prairie Creek drainage, was made obvious to me by an exercise I conducted in January 2004. I had six females in the Bald Hills that year with GPS and VHS transmitters, collared elk (Fig. 2.4). There was one person locating every collared elk during the same time that two to three other people (including me) were conducting surveys. The locator of collared elk did not conduct surveys or communicate findings over the days the rest of us were conducting surveys. Across the nine surveys that were conducted that year we sighted 34 female groups, of which 25 had one or more collared females. In the 25 groups, we only spotted the collared elk 67 percent of the time. I do not know whether we missed and did not count collared elk, or if we counted collared elk but failed to detect the collar. Nonetheless, these findings did not bolster my confidence that I counted every animal in groups or that I could readily pick out marked elk or naturally marked elk in the Bald Hills. Detecting naturally marked elk

FIGURE 2.5. An example of a naturally marked female elk from the Boyes herd. The female in front had an injury or deformity in the right eye. The two females in the back are examples of elk with normal eyes.

can require unobstructed views of stationary or slow-moving elk at distances of 10–200 m. Natural marks on elk can be subtle and not very apparent (Fig. 2.5). In comparison, it was not unusual in Prairie Creek to have to wait for elk to turn around so I could view a side I initially could not see; sometimes I had to move to view another part of the elk. Moving to get different views of elk was a luxury not possible in Bald Hills much of the time.

Another issue in Bald Hills was elk leaving the RNSP and using private lands in nearby areas (Kolbe & Weckerly, 2015). When elk left the Bald Hills they often went to Stover's Ranch, which was southeast of Bald Hills. I had permission to survey the meadows on the ranch. Yet, the meadows on Stover's Ranch extended more than 4 km from RNSP boundaries. Surveying this extended area as thoroughly as the Bald Hills was not possible given my shoestring budget. Consequently, it was possible that I did not detect elk in groups that might have used the ranch during the time I conducted my surveys in January.

ESTIMATES OR COUNTS OF ABUNDANCE?

The science of estimating population abundance can be quite involved (Williams et al., 2002). Nonetheless, conceptualizing the issues with estimating abundance is ($\hat{N}$) straightforward and boils down to dividing the count (C) by an estimate of count error, the sighting probability ($\hat{s}$);

$$\widehat{N} = \frac{C}{\hat{s}}. \quad 2.3$$

A count is simply the number of elk encountered; a sighting probability ranges from 0 to 1.0. For example, when every elk in a population is sighted only once during a survey then $\hat{s}$ equals 1.0 and the count is an enumeration of the population or census. Usually a census is not achieved during a count, so C is less than the true N and $\hat{s}$ is less than 1.0. One of the main reasons I strived to conduct 10 surveys in the Bald Hills and Prairie Creek each year was so that, hopefully, the highest count would be close to the actual abundance of the population, or what I call the high count.

Across a series of years, if the high counts are a similar proportion of the actual population size then $\hat{s}$ is similar across years and the condition of constant proportionality is met. I was a dinosaur programmed to think that if constant proportionality was met then high counts, albeit biased to some extent, would allow me to detect the extent of changes in abundance of the population across time. Constant proportion was, therefore, the essential condition necessary to quantifying variation in population dynamics.

Constant proportionality, however, can be relaxed. Johnson (2008) pointed out that high counts can signal changes in population abundance when the variance among the counts is greater than the variance among sighting probabilities. Another way to look at it is when sighting probabilities are high (e.g., 0.80–0.95) there is less opportunity for variation in $\hat{s}$ than there is in C (Weckerly, 2007).

To explore the feasibility that high counts signal abundance in each area, I analyzed Bald Hills and Prairie Creek survey data separately using different techniques. In the Bald Hills both missing elk and multiple counts of the same elk were real possibilities in a survey. In Prairie Creek where elk were habituated to anthropogenic activities, I assumed that double or multiple counts of the same elk did not occur. Hereafter, high counts are called counts. When describing the temporal patterns to a time series of high counts it can become confusing when, for example, there are a series of years with high counts that are low.

I estimated GSS models in the Bald Hills count data (Dennis et al., 2006; Hatfield et al., 2012; Seavy et al., 2009). State-space models were attractive because I could estimate abundance, σ^2—process variance—and τ^2—observer error. Observer error is also a variance; it estimates distortions in population change from imprecision of the survey technique. Imprecision in counts from a survey technique could be due to missing elk, counting the same elk multiple times, or both. Another more technical way to describe observer error is the error due to the sampling process and from measurement error. I viewed process variance as the signal and observer error as the noise.

GSS models also estimate *a*—exponential population growth, which would occur when abundance is low and density dependence is not expressed, and *c*—strength of density dependence. An appealing feature of the GSS model was the ability to estimate carrying capacity from estimates of *a*, *c*, σ^2, and τ^2. The carrying capacity calculated by the GSS model is the long-term mean of a stationary distribution of population abundances. In other words, population abundances that fluctuate around a mean carrying capacity, which I treated as *K* carrying capacity. I estimated parameters of GSS models using Bayesian statistics in the RJAGS packages in program R (R Core Team, 2014). When reporting parameter estimates of a GSS model, I also present the 95 percent credible intervals. I conducted three independent chains consisting of 1,000,000 iterations with an initial burn-in of 50,000 iterations and a thinning rate of 100. Convergence on all parameter estimates was obtained when the Brooks and Gelman diagnostic was not greater than 1.01 (Brooks & Gelman, 1998).

In Prairie Creek I estimated sighting probabilities and assessed potential factors that might influence elk sightings. Previous work suggested that sighting probabilities were high and factors that might influence sighting probabilities were complex (Weckerly, 1996, 2007). Factors that probably influenced elk sightings in the Prairie Creek drainage were group size, sex, whether or not the Boyes Meadows was burned, and elk abundance that year. When elk are in large groups they must spend more time in meadows, where they are readily sighted, because meadows contain most of the forage in the landscape. Sex might also influence sighting because males should have a broader forage niche than females (McCullough, 1999). In the Prairie Creek drainage that could mean males use forest more often, where they are harder to sight (Bliss & Weckerly, 2016). In January 1997, 2001, 2007, 2009, and 2014, much or all of Boyes Meadows (> 66 percent) had been burned the prior September or October. The prescribed fire was conducted shortly before the wet season began and the start of autumn green up. The burned away dead plant material allowed

TABLE 2.1. Findings from an AIC_C model selection analysis examining the influence of natural logarithm of typical group size (GS), sex (males were the reference category and coded 0, females were coded 1), whether Boyes Meadows was burned (burn), and high count (C) on sighting probabilities of naturally marked elk in the Prairie Creek drainage. The X denotes an interaction between two predictors. The selected model has an asterisk and its parameter estimates (coefficients) are presented below in logit scale. The delta refers to the difference in AIC_C between a model and the model with the smallest AIC_C. The number of parameters estimated by the model is nPar.

MODEL PREDICTORS	Δ	NPAR	DEVIANCE
*GS, sex, burn	0.0	4	1117.7
GS, sex	9.1	3	1128.9
GS, burn	22.0	3	1141.9
GS	41.4	2	1163.4
C, sex, burn	75.4	4	1193.2
Sex	78.1	2	1200.1
C, sex	80.1	3	1200.0
C, sex, C X sex	81.1	4	1198.8
Burn	137.1	2	1258.8
C, burn	139.0	3	1258.7
C	155.1	2	1277.1

COEFFICIENT	ESTIMATE	STANDARD ERROR
Intercept	−0.342	0.161
Typical group size	0.587	0.070
Sex	1.068	0.236
Burn	2.357	1.024

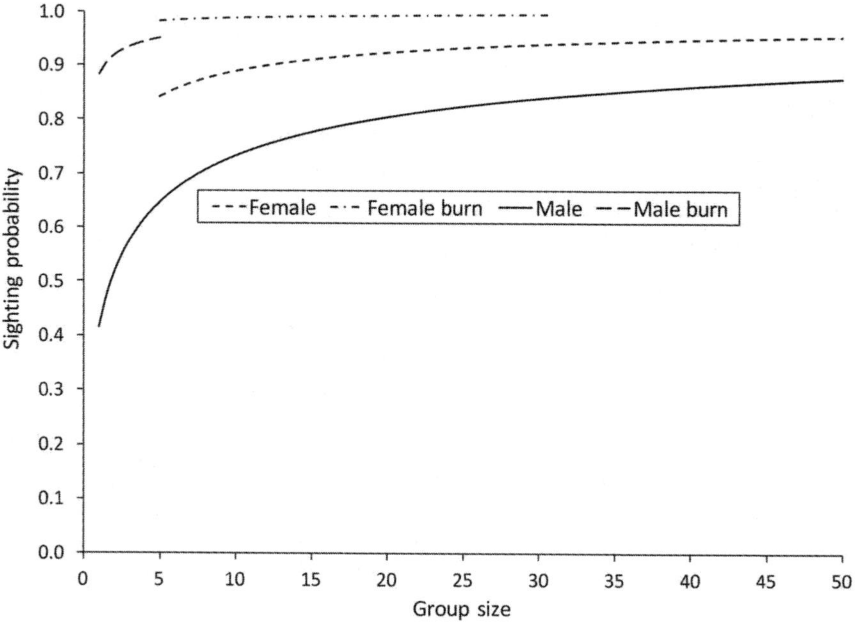

FIGURE 2.6. Estimated relationships between typical group size, sex, and burns on sighting probabilities of Roosevelt elk inhabiting the Prairie Creek drainage.

greater sunlight to the soil surface, conditions that probably favored plant growth and might have resulted in more green feed or easier access to feed for elk in January. Elk that used Boyes Meadows might then forage in meadows more frequently where they should be readily sighted during surveys. I also consider abundance of elk, indexed by counts, in the Prairie Creek drainage (males and females of all ages). Weckerly (2007) reported that the estimated abundance of females and males of all ages in the Prairie Creek drainage did not influence female sighting but that this measure of abundance was inversely associated with male sightings.

In Prairie Creek sighting probabilities were estimated from naturally marked (marked) elk using procedures described in Weckerly (2007) for a closed population. The data set consisted of sighting frequencies of 125 marked elk gathered between 1997 and 2015, 89 males and 36 females. I assumed that marked elk were a random sample taken from all elk inhabiting the Prairie Creek drainage (Weckerly, 2007). To reduce redundancy, I randomly selected one animal for analyses when two or more marked elk were together in one group. Marked animals sighted in the same group one or more times across the surveys conducted that year had identical or similar sighting frequencies, providing little new information to the data set. Typical group size was my measure of group size. The natural logarithm of typical group size was used because a preliminary analysis suggested a nonlinear, threshold relationship with sighting probability. A nonlinear, threshold relationship makes sense. At small group sizes elk can forage for some time in forests. But as group size increases the animals in the group must spend more time foraging in meadows. At a threshold size, larger elk groups must spend most of the time foraging in meadows because forest does not contain enough forage for all the mouths in the group.

My interpretation of the model selection analysis suggested that sighting probabilities were influenced by typical group size, sex, and whether or not most or all of Boyes Meadows had been burned the previous autumn (Table 2.1). Typical group size and burning positively influenced sighting probabilities, and females had higher sighting probabilities than males (Fig. 2.6).

The threshold for obvious change in sighting probabilities with typical group size depended on sex and whether elk used Boyes Meadows when it was burned or not burned. Female sighting probabilities were high (> 0.84) to very high (0.98–0.99), such as when females used Boyes Meadows in years when it was burned (Fig. 2.6). Male sighting probabilities could also be high and comparable to female sighting probabilities when they used Boyes Meadows in a year that it was burned. Males were either in male-only groups or in female groups (could be called mixed-sex groups). Groups comprised of only adult males were 20 or less in size. Male groups in this size range had sighting probabilities less than females.

A previous model selection analysis indicated that population abundance was uncorrelated to female sighting probabilities but inversely associated with male sighting probabilities (Weckerly, 2007). That analysis used data gathered in the Prairie Creek drainage during the first 10 years of the study (1997–2006). The present model selection analysis was not consistent with that finding. In the prior analysis, group size was not considered, and during the first 10 years, typical size of groups comprised entirely of adult males was correlated with population abundance ($r = 0.71$, degrees of freedom (df) = 8, $P = 0.021$). Yet, no correlation was detected between these two variables in the remaining 9 years from 2007 to 2015 ($r = 0.11$, $df = 7$, $P = 0.762$). Another difference between the present and previous analyses was that burning of the Boyes Meadows was not considered. As such, the analysis of the present data set was collected over a longer period of time than the previous data set and therefore captured more environmental heterogeneity. Population abundance does not influence sighting probabilities as strongly as does group size.

The nonlinear relationships between group size and sighting probability indicate that group size can be either a promoter or a dampener of sighting variability. This outcome depends on whether group sizes are below or above thresholds on sighting probability curves. To illustrate, I used data on all group sizes of females and males encountered across all surveys from 1997 to 2015 in the Prairie Creek drainage. Group sizes in this analysis were treated as proxies of counts. Female group sizes tended to be large: 77 percent of all female groups had more than 20 elk (Fig. 2.7). Group sizes greater than 20 were also above the threshold on sighting probability curves (Fig. 2.6). Using the nonburn, female curve as an example, the sighting probability was 0.92 when group size was 20 but the sighting probability increased only to 0.95 when group size was 50. There could be a two-fold difference in group size but the effect on sighting probabilities was miniscule. Female group size was a dampener of variances among sighting probabilities. On the other hand, male group sizes tended to be small; 68 percent of all male groups encountered had 10 or fewer animals (Fig. 2.7). Now using the nonburn, male curve as the example, a group size of 5 had a sighting probability of 0.65. When group size was 10, the sighting probability increased by 0.13 to 0.78. Male group size was a promoter of variance among sighting probabilities.

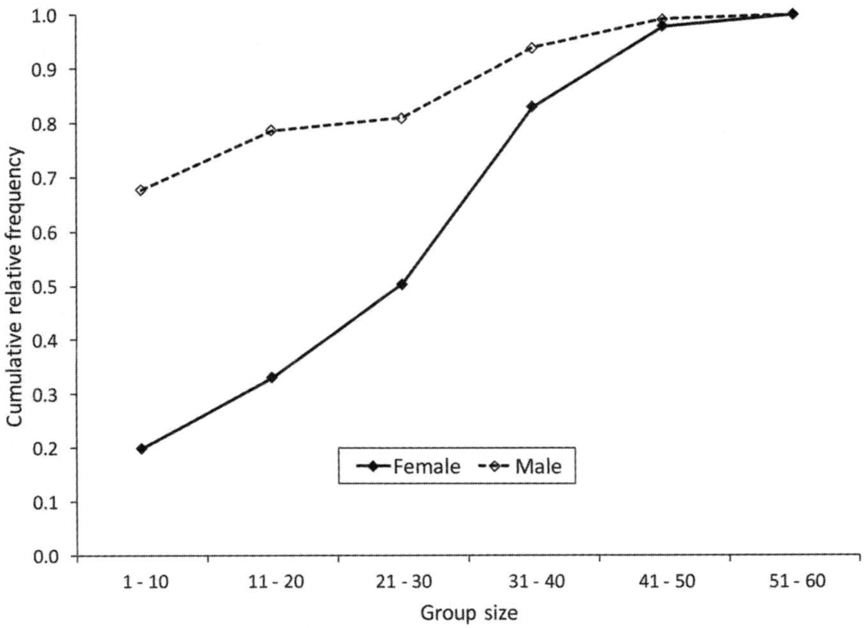

FIGURE 2.7. Cumulative relative frequency of all female and male groups encountered during population surveys along the Prairie Creek drainage, 1997–2015. Female groups were comprised of females, juveniles, subadult males, and sometimes a few adult males. Male groups were comprised of adult males only and the times when males were encountered in female groups. Sample sizes were 312 females and 647 males.

To compare variability of group sizes to variability of sighting probabilities, I calculated coefficients of variation (CV, standard deviation/mean) in group sizes and sighting probabilities. Between 1997 and 2015 I encountered 312 female groups and 647 male groups (Fig. 2.7). As a refresher, male groups were comprised entirely of males and female groups could have adult males but were mostly comprised of females, juveniles, and subadult males. Expected sighting probabilities were obtained from the selected model in Table 2.1. Again, sighting probabilities depended on group size, group type (male only and female groups), and whether animals used Boyes Meadows in a year when it was burned or not burned. For females, the CV of group size was 0.51 and 0.05 for sighting probability. The CV ratio (group size/CV/sighting probability CV) was 11.07. Variability in group size was 11 times greater than variability in sighting probabilities. For males, the coefficients of variation (CV) for group size was 1.20 and 0.25 for sighting probability. The CV ratio was 4.79. In other words, variability in group size was 4.79 times greater than that of sighting probabilities. It is also apparent that this ratio was considerably less than the female ratio. Count data for the female segment of the population signals actual abundance.

Another way to evaluate whether counts are a signal of abundance is to compare count data collected by two different sets of surveyors that do not communicate their findings to each other. Although the counts are obtained from the same population the two sets of counts are gathered independently of one another. There should be strong agreement in the two sets of counts in a setting with high sighting probabilities and where variability in sighting probabilities is markedly less than among counts. From 1997 to 2006 RNSP biologists also surveyed the two female herds in the Prairie Creek drainage, Boyes and Davison. The RNSP biologists conducted up to four surveys that were within one to four months of when I conducted my surveys. Because we did not conduct our surveys in the same months, there probably is some disagreement in counts because of elk mortality. Nevertheless, we did not communicate our findings to each other during these years. If there was agreement between counts, then the intercept coefficient should be zero and the slope coefficient should be 1.0 for the simple linear regression. The r^2 value, also, should approach 1.0.

There was strong agreement in counts (Fig. 2.8). The simple linear regression summarized data well ($r^2 = 0.94$). The slope estimate was similar to 1.0 but the estimate of the intercept was less than zero. The greater deviation between counts seemed to happen when herd abundance was low. The greater deviation in counts between myself and RNSP biologists might be due to low sighting probabilities of small group sizes when abundance was small.

Determining whether the Bald Hills counts could signal abundance could not rely on sighting probabilities that ranged from 0 to 1.0. Multiple counts of the same elk during a survey in the Bald Hills was a possibility and so the upper bound of a sighting probability could exceed 1.0, which technically is not a possibility. Estimating sighting probabilities was out of the question. I therefore estimated abundance, process variance, and observer error in the Bald Hills time series using the GSS model.

Credible intervals of abundance estimates obtained from the GSS model overlapped counts in 12 of 18 years (Fig. 2.9). The greatest deviation between abundance and count was in 2016 when the count of 180 was 64 elk fewer than the estimated abundance (244). Either there was a dramatic drop in abundance because of mortality or emigration, or I failed to sight a substantial part of the herd during surveys in 2016. Nonetheless, process variance and observer error

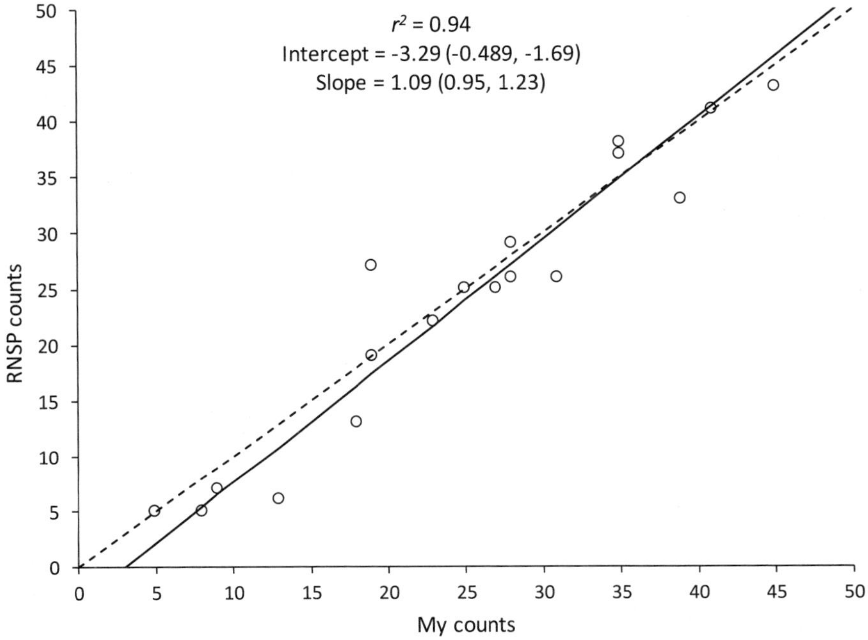

FIGURE 2.8. Scatterplot of high counts of the Boyes and Davison herds obtained by myself and by RNSP biologists from 1997 to 2006. The 95 percent CI for the slope and intercept are presented in parentheses. The dashed line is the 1:1 line. All data points would touch this line when there is complete agreement between counts. The estimates of regression coefficients of the 1:1 line would be an intercept of 0.0 and a slope of 1.0. I used bisector regression to estimate coefficients to accommodate the error in the predictor variable (Weckerly, 2007).

estimates were 0.00810 and 0.00314, respectively. The signal, estimated as process variance, was more than twice the noise, estimated as observer error, in the Bald Hills time series.

There are strengths and weaknesses to estimating abundance from observed counts or from abundance estimates that accommodate counting errors (Hunt, Weckerly, & Ott, 2012; Weckerly, 2007). An abundance estimate that is reliable will be precise and unbiased (Weckerly & Foster, 2010). Indeed, the great appeal to using an abundance estimator is the possibility of unbiased and precise estimates. The primary drawback to population estimators is they require assumptions. In other words, an estimate is only as believable as its foundation—the assumptions of the estimator (Johnson, 2008). Whether assumptions are met for an estimate spans the spectrum from a solid brick or cement foundation to a foundation that is quicksand. It can be difficult to assess

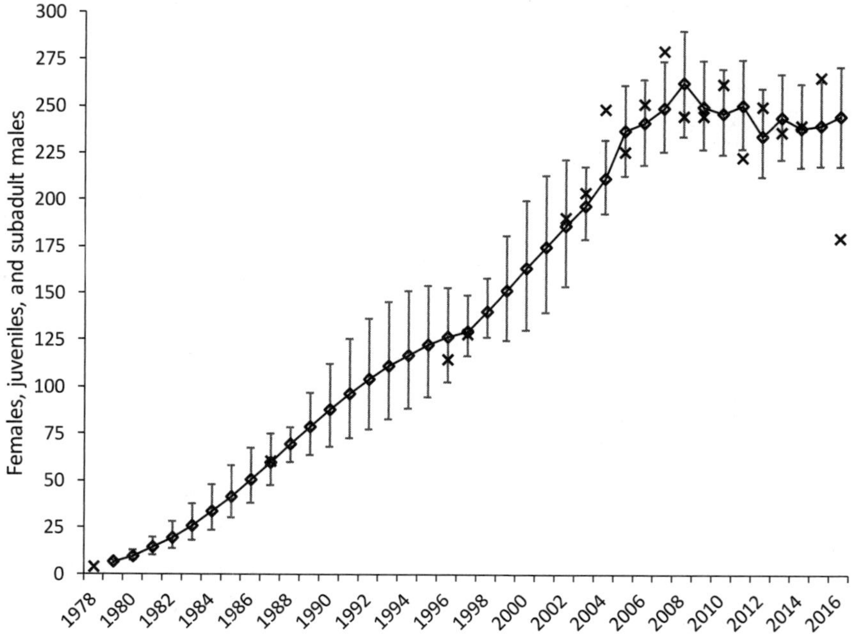

FIGURE 2.9. Counts of the Bald Hills herd between 1978 and 2016 (crosses) as well as abundance estimates (open diamonds), with 95 percent credible intervals. Abundance estimates were obtained from a GSS model analyzed with Bayesian statistics.

the material of the foundation. A common criticism of count data is that often there is no analysis of the extent that counting errors distort the utility of counts to signal abundance. Someone can argue that the noise swamps out the signal and there are no data to refute the argument. My analyses indicate that counts provide a strong signal of female abundance in the Prairie Creek drainage and a signal of female abundance in the Bald Hills population. Habituation of elk in the Prairie Creek drainage probably has something to do with the strong signal there. Elk do not flee from people so it is possible to observe them for the time it takes to count each and every animal in the group. Elk inhabiting the Gold Bluffs and areas around the town of Orick in the lower Redwood Creek drainage are also habituated to people (Fig. 1.5). These elk herds are counted by RNSP biologists three to eight times a year, each autumn (Schmidt, 2015). That survey protocol is probably sufficient to obtain count data that signals abundance. This is beneficial to me because I can include those data in my analyses and findings about resiliency and redundancy. When reporting and analyzing time series of herd abundances and herd attributes such as juvenile: female ratios, I use count data.

Conducting the population surveys was as much fun the first time as it was the last time, 20 years later. Doing the population surveys was also a marker of time. In the beginning, some students struggled to keep up with me on hikes. In the end, I had trouble keeping up with many graduate students on hikes. As enjoyable as it was collecting count data, the meaning of the count data was more important. The next four chapters present my findings and interpretations to understand the dynamics of herd distribution and abundance.

3

Distribution

Population dynamics of the nonmigratory Roosevelt elk are influenced by their social organization and the manner in which they make use of their living space.

— Franklin, Mossman, & Dole,
"Social Organization and Home Range of Roosevelt Elk"

The distribution of a population needs to be delineated to observe, measure, and record something about the herd. Open habitat is strongly tied to the distribution of individual elk in the forested landscapes of north-coastal California (Franklin et al., 1975; Harper et al., 1967; Weckerly, 2005). Open habitat essentially consists of two types, meadow and open forests where the trees have recently been removed by logging or burned by fire. Because timber harvesting and forest fires have not occurred in the RNSP for roughly 40 years, what was open forests at one time is now forest with closed canopies and little elk forage (Weckerly, 2005). In the landscape of the RNSP during the time of this study, elk distribution is clustered in meadows that have the bulk of forage and in forests that are near meadows.

FACTORS LIKELY INFLUENCING FOOD INTAKE

The cues that animals use to make decisions about where to forage and for how long are vital to understanding distribution (Merkle, Fortin, & Morales, 2014; Wolf et al., 2009). In January food in meadows is not as abundant or as nutritious as it will be in springtime (Fig. 0.3). But it is green. There appear to be some nutritional rewards from foraging decisions at this time of year, because month-to-month diet quality no longer seems to decline, as it does from summer to late autumn (Fig. 0.3). Furthermore, the foraging decisions elk make in January influence survival and production because of the complexity of the ruminant digestive system (Barboza et al., 2009). Integral to digestion of plant material is maintaining the environment necessary for fermentation by bacteria, fungi, and protozoa in the gut. Digestion by elk ceases when gut microorganisms do not have a constant input of substrate (i.e., plant material that can be digested).

Forage intake is the product of bite rate and bite size. Measuring bite rate in free-ranging elk can be done, but measuring bite size is difficult. I tried and failed miserably. The miserable part was the time and effort devoted to measuring bite size that, in the end, appeared to result in biased estimates of the forage ingested in a bite. Therefore, I had to resort to extracting data from studies that directly measured forage biomass (kg dried biomass $\cdot$ ha^{-1}) and bite size using tame elk (Fig. 3.1). Bite size displays a positive, nonlinear relationship with forage biomass. At low forage biomasses, there is a noticeable increase in bite size with biomass, and the relationship begins to flatten at about 1,000 kg $\cdot$ ha^{-1}. The relationship was evident because the fixed factor, biomass, accounted for 55 percent of the variation in bite size (marginal r^2). There was variability among studies from which I extracted the data (conditional r^2 0.28 greater than marginal r^2). Variability across studies was probably because of differing techniques used to measure bite size and study goals. Assuming that free-ranging Roosevelt elk take bites like tame elk, I used my forage-biomass estimates in meadows to then estimate bite size from the biomass–bite size regression in Figure 3.1.

Using parts of meadows probably is dictated by at least two factors. One, of course, is the biomass of the food supply. The other factor is the time spent grazing when the animal is engaged in feeding activity. Elk grazing is when the head is at ground level taking bites or below the shoulders in which case the animal is probably searching for the next bite. I call grazing "head in the feeding position" (HFP). In January, when green grass and forbs in meadows are short in height and forage biomass is low, HFP might have substantial influence on food intake.

The relationship between HFP and bite rate, however, is not necessarily straightforward. Chewing and cropping, or taking a bite, are competing processes across the spectrum of scarce to abundant forage biomass. All else being equal, when bite sizes are large, animals have to chew to break down material into smaller particles before they swallow them. An inverse relationship between bite and chew rates, therefore, is expected and has been documented

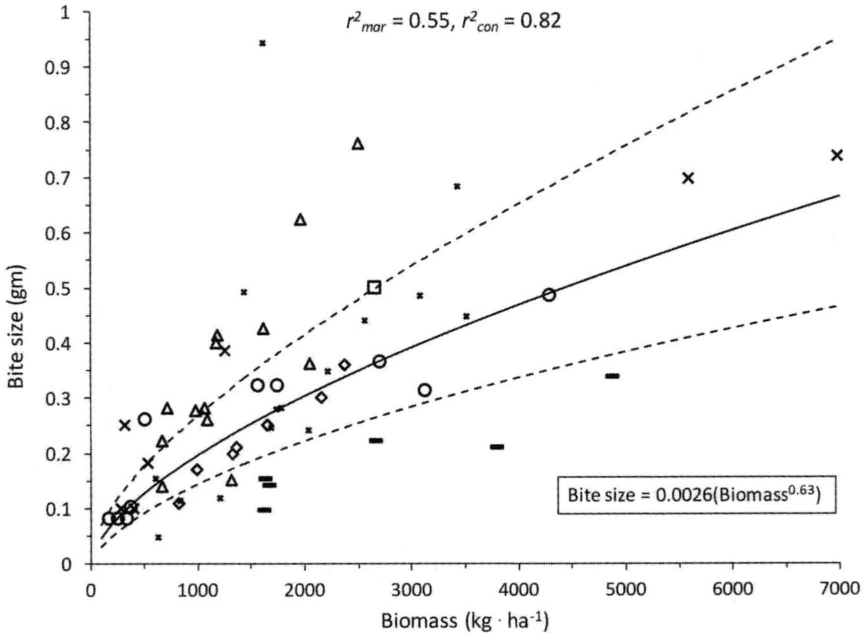

FIGURE 3.1. Scatterplot and regression (along with 95 percent confidence envelope) of forage biomass and bite size. Forage biomass was grass or vegetation in grasslands. The regression was estimated with a mixed effects model with study as the random factor. Data were obtained from Arthur, Hudson, Donkar, and King (2009, triangle), Collins, Urness, and Austin (1978, square), Gedir and Hudson (2000, bar), Hudson and Nietfeld (1985, diamond), Hudson and Watkins (1986, circle), Wickstrom, Robbins, Hanley, Spalinger, and Parish (1984, large x), and Wilmshurst et al. (1995, small x with vertical bar). The 95 percent CI of the slope estimate was 0.52–0.73. Also reported are the marginal and conditional r^2 values.

(Gross, Shipley, Hobbs, Spalinger, & Wunder, 1993). In Davison Meadows forage biomass is no greater than 500 kg · ha^{-1} most of the time in January (Fig. 3.2). Forage biomass is on the scarce end of the forage biomass spectrum. From 100 to 500 kg · ha^{-1} of forage, not only does bite size increase steeply with increased forage biomass, but the precision of estimated bite sizes is also greater. I would not expect chewing and cropping to compete across this relatively narrow range of forage biomasses because bite sizes have to be small in size. This appears to be the case, because there was little hint of an inverse, or any, relationship between chewing rate and bite rate (Fig. 3.3). If chewing does not impede cropping, then a positive relationship between HFP and number of bites taken is possible. It is also plausible that elk could take more bites in areas with more forage biomass to increase food intake in January. When chewing

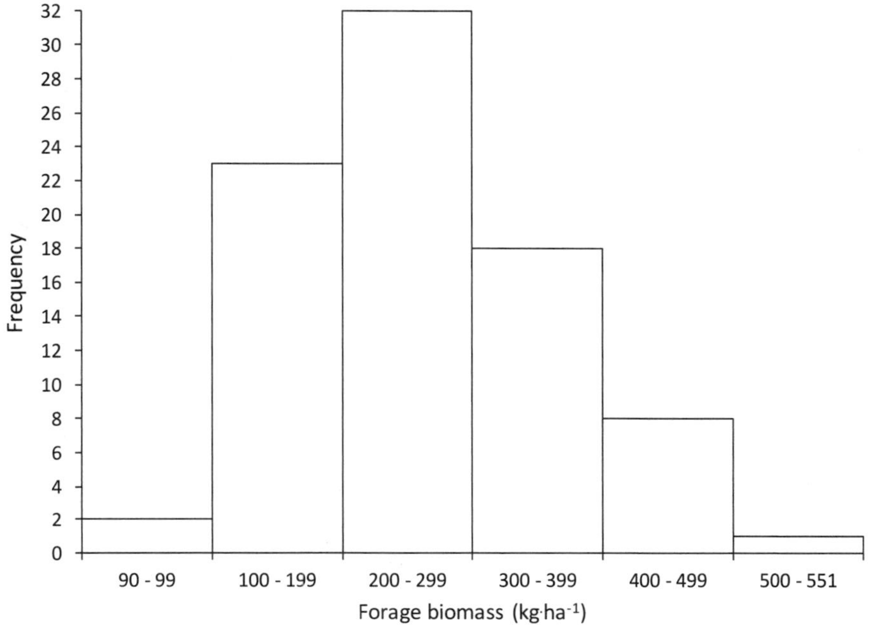

FIGURE 3.2. Histogram of estimated forage biomass in each of the seven sectors in Davison Meadows for each year between 2005 and 2016.

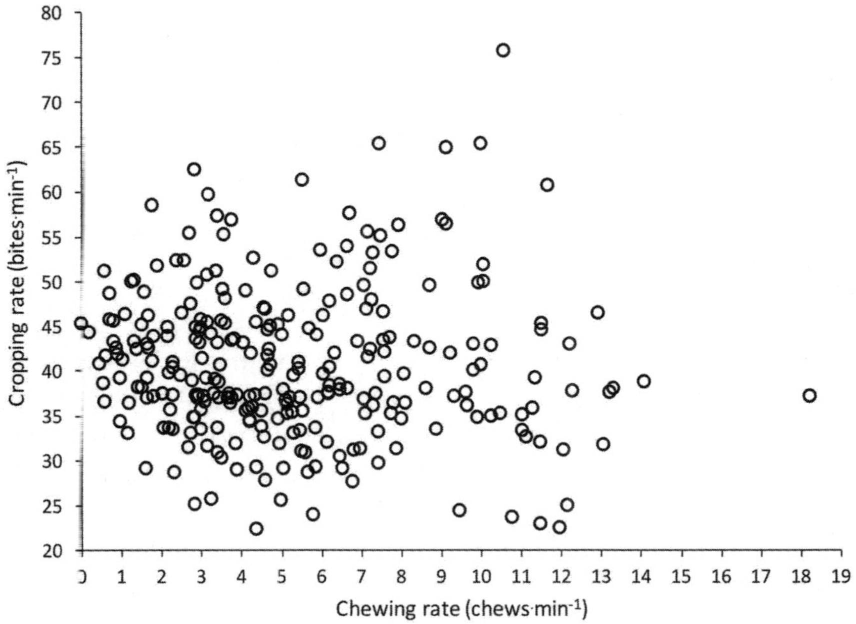

FIGURE 3.3. Scatterplot of chewing and cropping rates (per minute) revealing no influence of chewing on cropping. Data were obtained from focal animal observations of elk in the Prairie Creek drainage. A mixed effects model revealed no influence of chewing rate on bite rate ($F = 0.008$, $df = 1, 257.3$, $P = 0.929$). The influential fixed factors in the model were forage biomass in the sector ($F = 15.23$, $df = 1, 160.8$, $P < 0.001$) and group size ($F = 14.7$, $df = 1, 189.6$, $P < 0.001$). The random factor was sector.

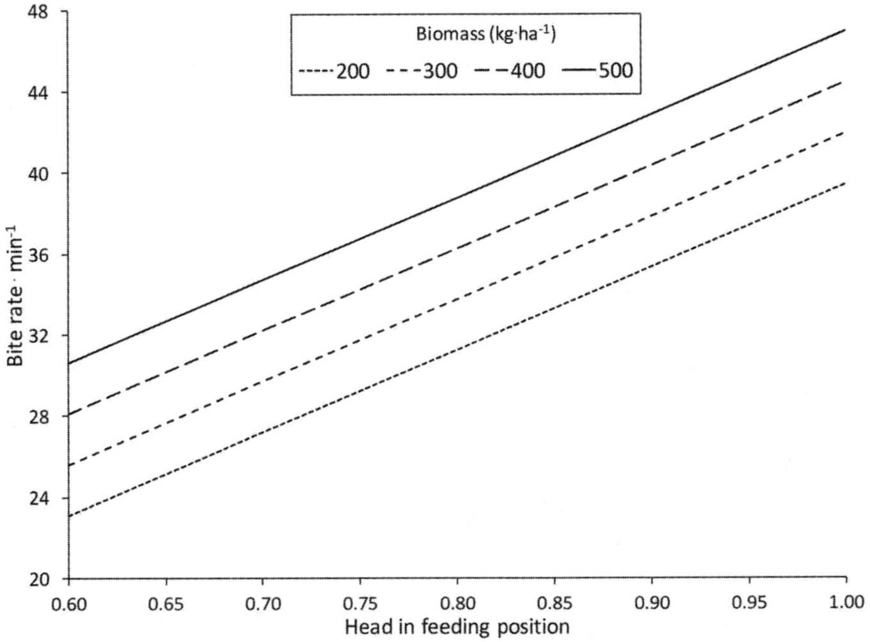

FIGURE 3.4. Estimated relationships between head in the feeding position (proportion of a minute), forage biomass, and bite rate. Bite rate (number of bites taken in a minute) relationships were estimated from a mixed effects model that had fixed predictors of head in the feeding position ($F = 159.2$, $df = 1, 257.3$, $P < 0.001$), forage biomass in the sector ($F = 15.4$, $df = 1, 184.4$, $P < 0.001$), and group size ($F = 15.7$, $df = 1, 201.4$, $P < 0.001$). Sector was the random factor. The marginal and conditional r^2 values were 0.40 and 0.64, respectively. The relationships were estimated using the mean group size of the data (25). The regression (SE) was as follows: Bite rate = −2.76(3.74) + 40.8(3.24) × head in feeding position + 0.025(0.006) × biomass−0.15(0.04) × group size. Sample size was 271. The biomasses chosen (200 to 500 in increments of 100) to be plotted reflect the variety of biomasses available to elk in January.

impedes the time an elk has to take bites, there is an inverse relationship between forage biomass and bite rate (Wilmshurst, Fryxell, & Hudson, 1995; Weckerly et al., 2001). When forage biomass is high, elk take larger bites that require more time to chew. The consequence is a lower bite rate and possibly more time for the elk to have the head up while chewing and swallowing material in the mouth. Fortunately for me, it is credible to expect a positive relationship between HFP, forage biomass, and bite rate in January because chewing and cropping are not competing processes for elk.

Indeed, forage biomass and HFP were positively related to bite rates (bites · min^{-1}) in January (Fig. 3.4). Elk take more bites in places with greater feed and

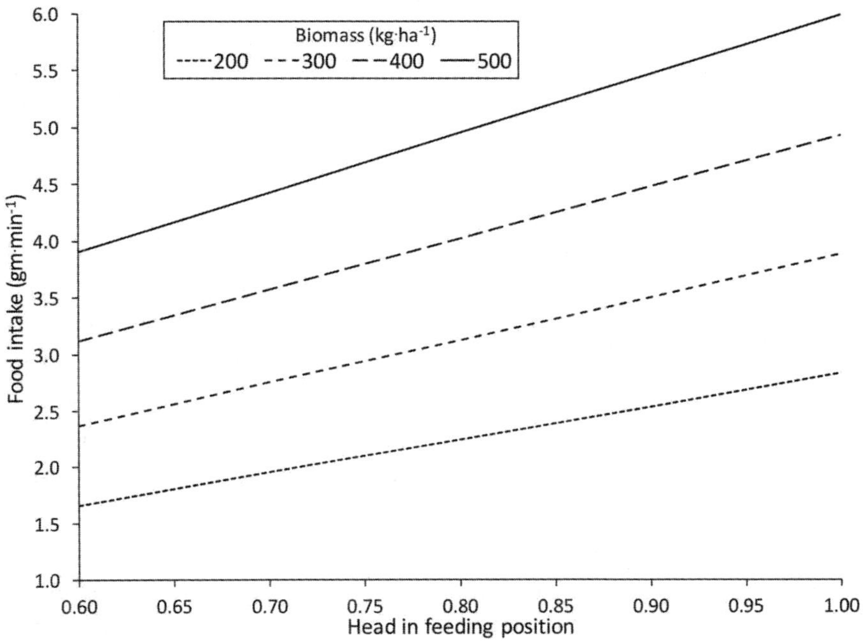

FIGURE 3.5. Expected food intake (bite size × bite rate) in relation to the proportion of one minute the head is in the feeding position and forage biomass. Bite sizes were obtained from regression reported in Figure 3.1, and bite rate was obtained from regression reported in Figure 3.4.

when HFP is longer. Since food intake is the product of bite rate and bite size, I used the regressions presented in Figures 3.1 and 3.4 to project food intake. Food intake increased with greater forage biomass and when HFP increased from 0.6 (36 secs) to 1.0 mins (Fig. 3.5). HFP was between 0.6 and 1.0 in 98 percent of all focal observations. What intrigued me was the potential impact of HFP on food intake. For example, if an elk fed in a place with 300 kg · ha^{-1} of feed with an HFP of 0.6, projected food intake was 2.4 gm · min^{-1}. If, on the other hand, an elk fed in a place with 200 kg · ha^{-1}, the same food intake would be achieved with an HFP of 0.85. Adjusting HFP to maintain or increase food intake in places with low forage biomass is a plausible possibility in January. Also to be considered, animals that have a lower HFP in some places might be reluctant to go to those places to feed because of reduced food intake.

Elk decision making about where to feed is influenced by forage biomass, but might also be affected by disturbance and perceived risks from people and natural predators (Bowyer et al., 1998). Obviously, elk should feed in places with more forage biomass for a longer period of time because of elevated food intake. Conversely, in places where elk are constantly disturbed, the greater vigilance would result in a lower HFP. Elk might then decide to feed elsewhere. Disturbance could be from puma and black bear. Disturbance also could be from people that get too close to elk or when someone's dog chases elk.

MEADOW PARTITIONING

Understanding something about the spatial distribution of herds can be simple or complex. The Gold Bluffs herd uses small meadows and beach grass habitat clustered along the coast north of Orick (Bowyer, 1981; Franklin & Lieb, 1979; Franklin et al., 1975). This area is separated by 3–5 km of forest from meadows where the nearest herds occur in the Prairie Creek and lower Redwood Creek drainages (Fig. 1.5). What determines the boundaries of the Gold Bluff herd seems straightforward: it is the ocean to the west and the forest just beyond the cluster of small meadows and beach grass habitat in the remaining cardinal directions. For the other herds in the Prairie Creek and lower Redwood Creek drainages, spatial boundaries involve using meadows that also are used by at least one other herd, which results in meadow partitioning. Understanding the mechanisms that maintain the spatial boundaries of meadows partitioned by different herds is helpful to understanding the observed number of herds in this region of the RNSP, the dynamics of boundaries, and population dynamics.

In January forage biomass in meadows should be more limited than in late winter through early summer (Fig. 0.3). At the same time, forage is distributed heterogeneously within meadows (Fig. 3.2). Places in meadows with less forage should be foraged less than places with more forage. Elk also should move among sectors (Fig. 2.1) or different parts of Davison Meadows to maintain spatial memory of food supplies (Wolf et al., 2009). There should be a linear relationship between food supply and use, with an intercept of 0 and a slope of 1.0 if there are not complications that could distort that relationship. Potential complications are from the elk themselves as well as from predators and people. Adult males are socially dominant to females and might influence herd spatial patterns. Predators might pose greater lethal risks to elk in particular parts of meadows. Poaching might have the same effect. To my knowledge, how availability–use relationships are affected by two herds partitioning meadows has not been examined in much detail.

I estimated food supply–elk forage use in the Davison Meadows in January between 2009 and 2012. Early in my study (1997) females and males were partitioning parts of Davison Meadows (Weckerly, 1998; Weckerly, McFarland, Ricca, & Meyer, 2004). Intersexual partitioning appeared to be associated with female groups avoiding large groups of males (> 6, Weckerly et al., 2001). Meadow partitioning between females and males, however, was not evident when data on forage activity were collected between 2009 and 2012. Group segregation is a likely result when females avoid male groups that tend to be large. Sexual segregation and aggregation coefficients (SSAC) were calculated to assess whether males and females segregated, aggregated, or associated randomly (Bonenfant et al., 2007). I calculated the SSAC from data collected on solitary elk and groups during population surveys conducted in the Prairie Creek drainage. The SSAC ranges between 0 and 1.0. From the same data, 10,000 random permutations were generated to calculate a 95 percent confidence band (CB) from the permuted SSAC values. The observed SSAC in relation to whether it was above, below, or within the 95 percent CB determined whether males and females were, respectively, sexually segregated, aggregated, or randomly associated. In 1997 females and males within Davison Meadows were sexually segregated (SSAC = 0.89, 95 percent confidence band: 0.48–0.81, $n = 48$ groups and solitary animals). The typical size of groups comprised entirely of males was 13 in 1997 (bootstrapped 95 percent CI: 10–14, $n = 21$). In contrast, from 2009 to 2012 females and males in Davison Meadows were randomly associated (2009: SSAC = 0.79, 95 percent CB: 0.50–0.83, $n = 49$; 2010: 0.54, 0.43–0.78, $n = 37$; 2011: 0.71, 0.49–0.85, $n = 45$; 2012: 0.70, 0.37–0.78, $n = 26$). Typical size of male groups during those years was 3 (95 percent CI: 2–3, $n = 50$ groups). The Davison herds foraging time in sectors of Davison Meadows from 2009 to 2012 did not appear to be influenced by large groups of males because there were none.

Elk might perceive risks differently among sectors, which might influence time that elk feed (Bowyer et al., 1998). To assess differential risk, I measured time that the head was up when elk were foraging, which is my measure of vigilance. Vigilance is when the head is above the shoulders and is a behavior displayed by elk when scanning for other elk, predators, or to assess potential risks from RNSP visitors (Stevens, 1965). Focal animal observations of 4–10 mins in length were collected in six of the seven sectors in Davison Meadows between 2009 and 2012 (Table 3.1). Elk vigilance was similar across sectors in Davison Meadows. By extension, risk and disturbance were also similar across sectors.

TABLE 3.1. Parameter estimates (logit scale) of two linearized mixed effects models with binomial error structure to estimate vigilance (seconds head up length of observation in seconds^{-1}) by elk in the Davison herd. Both the complete and reduced models had predictors that influenced time the head was up when grazing. The predictors were age–sex categories (females, juveniles, and subadult males), proportion of time another elk was within one body length of focal elk (one body length), forage biomass (kilograms · hectare^{-1}), and length of the focal observation (minutes). The complete model had additional predictors for the sector where the elk was measured. Random effects for both models were the focal observations. If elk were perceiving risks from park visitors and natural predators differentially across sectors, then I expected the complete model to summarize the data better than the reduced model. Findings from a likelihood ratio test of the two nested models, however, indicated the reduced model summarized the data well enough ($\chi^2 = 2.9$, $df = 5$, $P = 0.716$). The log likelihood of the complete model was −1458.9 and for the reduced model it was −1460.3. Reference categories were sector A and juveniles. Total sample size was 285, with 27 to 52 observations in each of the six sectors.

	COMPLETE		REDUCED	
COEFFICIENTS	ESTIMATE	SE	ESTIMATE	SE
Intercept	−2.07	0.31	−2.31	0.18
Sector				
Sodav	−0.34	0.41	—	—
B	0.06	0.43	—	—
C	−0.38	0.30	—	—
WPC	−0.29	0.44	—	—
HB	−0.23	0.48	—	—
Female	−0.48	0.23	−0.46	0.23
Subadult male	−0.25	0.30	−0.28	0.30
One body length	−0.28	0.10	0.28	0.10
Forage biomass	−0.17	0.16	−0.20	0.10
Minutes	−0.29	0.11	−0.28	0.10

AVAILABILITY AND USE

I measured foraging activity by the Davison herd in January 2009–2012, in the sectors of Davison Meadows (Fig. 2.2). The Levee herd also foraged in the Davison Meadows. Forage intensity of the Davison herd was measured by the number of elk and the hours the elk fed (elk hours) during daylight hours. Because of restrictions from the shoestring budget, I was unable to measure elk activity at night, a time when elk feed (Bowyer, 1981). I also did not have the personnel to measure elk hours that the Levee herd fed, but I was able to determine the number of hours this herd fed in the sectors of Davison Meadows. Identifying different herds is based on four things (Weckerly, 1999): (1) location of herds within an hour of one another, (2) marker elk, (3) group size, and (4) group composition. Marker elk were individuals with unique features that facilitated herd identification. I first noticed the Levee herd in Davison Meadows in 2007. The Levee herd was located in the south end of the Davison Meadows (i.e., sectors sodav and C in Fig. 2.2) in 2009, 2010, and 2012. In 2011 the Levee animals also ventured to the north end of Davison Meadows, but this herd spent 78 percent of the time in sectors sodav and C. I conducted a model selection analysis to determine the factors that influenced the relationships between the food supply and elk foraging by the Davison herd. Models were built to assess the following questions:

- Did sector food supplies (kilograms · hectare^{-1} × number of hectares) influence elk foraging?
- Did sector food supplies influence elk foraging, or did the food supplies in a sector and all adjacent sectors influence food supply–foraging relationships?
- Did the food supply–elk foraging relationships differ across years for reasons I did not understand or appreciate?
- Did the presence of the Levee herd, or the time that herd spent, in the south end of Davison Meadows (sectors sodav and C) influence food supply–foraging relationships?
- Finally, did the presence or absence of the Levee herd in every sector visited, or the time spent in sectors, influence food supply–foraging relationships?

There undoubtedly could have been more models considered. Number of parameters to consider in models, however, was constrained by sample size (7 sectors × 4 years = 28 data points).

The response variable was elk hours the Davison herd fed in sectors, expressed as a proportion of total elk hours that year (Table 3.2). Food supplies in sectors were also expressed as a proportion of total forage biomass in Davison Meadows that year. These two variables were converted to proportions because elk forage use should be in regard to food supplies that year. I estimated linear mixed effects models. The model selection summary indicated that there were three competing models ($\Delta < 2.0$). Nonetheless, I selected the model with forage biomass in the sector (FEED) and whether or not the Levee herd was present in sectors sodav and C (PRES) as the predictors. This model had a somewhat higher log likelihood than the model with only FEED as the predictor and a slightly higher log likelihood than the model with predictors of FEED and hours that the Levee herd was in sectors sodav and C (LEVHRS). The marginal r^2 of the selected model was also the highest of any model considered. I also report the parameter estimates of the competing models to show that, regardless of the model, the slope estimates for FEED did not differ from 1.0.

TABLE 3.2. Findings from an AIC_c analysis summarizing the proportion of elk hours feeding in Davison Meadows sectors. Models were linear mixed effects models with sector as the random factor. The selected model has an asterisk. Below the model selection summary are the parameter estimates of the selected model and summary of a test revealing that the slope of FEED was similar to 1.0. Also presented are the parameter estimates of the two models with Δ less than 2.0 as well as test findings revealing that the slope of FEED was similar to 1.0. Elk hours was the sum of the number of elk and the hours each elk in the Davison herd fed in a sector that year. Elk hours were expressed as a proportion of total elk hours for that year. The model selection summary also includes marginal (r^2_{mar}) and conditional r-squared (r^2_{con}) values for each model. Predictors were: FEED—forage biomass in sector (kg) expressed as a proportion of total forage biomass in Davison Meadows that year; AREA—size of sector; ADJ—forage biomass in all adjacent sectors expressed as proportion; YR—categorical variable for year; PRES—presence (reference category) or absence of Levee group in the partition zone (sectors sodav, C) that year; LEVHRS—hours the Levee group was in sectors sodav or C that year expressed as proportion; LEVALL—presence or absence of Levee group in all Davison sectors; LEVALLHRS—hours Levee group was in any Davison sector that year expressed as proportion; NO—no fixed predictors.

PREDICTORS	LL	Δ	NPAR	R^2_{MAR}	R^2_{CON}
*FEED, PRES	30.28	0.0	5	0.49	0.55
FEED	28.51	0.5	4	0.40	0.61
FEED, LEVHRS	29.60	1.4	5	0.48	0.53
FEED, LEVALLHRS	29.21	2.2	5	0.44	0.55
FEED, LEVALL	28.53	3.5	5	0.41	0.61
NO	24.44	6.0	3	0.00	0.52
FEED, AREA, ADJ	28.86	6.1	6	0.40	0.60
FEED, YR	28.51	10.4	7	0.40	0.61
FEED, YR, FEED X YR	30.31	20.2	10	0.46	0.68

Selected model

FIXED COEFFICIENTS	ESTIMATE	SE	DF	t	p
Intercept	−0.012	0.048	12.3	−0.23	0.797
FEED	1.435	0.441	22.0	3.25	0.004
PRES	−0.174	0.10	14.0	−1.75	0.103

t-test slope = 1.0: $t = 0.99$, $df = 22.0$, $P = 0.334$

Model: Δ = 0.5

Intercept	0.028	0.048	7.7	0.58	0.576
FEED	0.802	0.271	9.7	2.96	0.015

t-test slope = 1.0: $t = -0.73$, $df = 9.7$, $P = 0.481$

Model: Δ = 1.4

Intercept	0.028	0.041	4.8	0.65	0.542
FEED	0.888	0.250	7.1	3.55	0.009
LEVHRS	−0.104	0.080	22.0	−1.30	0.207

t-test slope = 1.0: $t = -0.45$, $df = 7.1$, $P = 0.668$

For every 1 percent change in food supply there was a 1 percent change in forage use. What I expected.

What I assessed but did not anticipate was an intercept changing as a function of whether or not the Levee herd was present in sectors sodav and C. It was plausible to disregard the PRES coefficient because the P value for the coefficient was greater than 0.05. The PRES coefficient would then be interpreted as statistically nonsignificant, meaning the intercepts of the regression of the Davison herd in sodav and C sectors and the regression of that herd in the remaining sectors to the north were the same. If you accept that interpretation, then select the model with a slightly larger AIC_C ($\Delta = 0.5$, Table 3.1). The intercept is similar to 0.0 as I expected. When there is no food there should be no feeding by elk, which seems intuitive. Be that as it may, when models are within two AIC units, nested, and differ by one predictor, I took the suggestion of Arnold (2010) and relaxed the cutoff for statistical significance to 0.15. The model with an additional predictor and a log likelihood somewhat larger than the next model will have $P < 0.157$, which is what I found. The Davison herd appeared to avoid the southern part of Davison Meadows (sectors sodav, C) that was used by the Levee herd most often (Fig. 3.6). In the parts of Davison Meadows to the north, the intercept was estimated to be similar to 0.0. In the south end of Davison Meadows, the presence of the Levee herd appeared to alter foraging intensity of the Davison herd. It is conceivable that social interactions between these two herds might affect the use–availability relationships I detected.

Yet, the relationships detected by interpreting the selected model were not strong. The fixed predictor accounted for 49 percent (marginal r^2) of the variation in foraging intensity by the Davison herd. Perhaps the Davison herd is more active at night and moves a lot. If I had measured feeding activity in the day and night, I might have detected stronger relationships. Yet, the Davison herd did not appear to move more at night than during the day. On 22 occasions when I left the herd at dark after following it all day, I located the herd the next morning at first light. When adjusted for number of hours between these locations, the straight-line distance between the morning and the evening locations was actually greater than the evening location and the herd's location the following morning (Fig. 3.7). During the day, mean straight-line distance traveled per hour was 51 m whereas during the night it was 27 m. In terms of activity the Davison herd spent 63 percent (SE = 0.5, $n = 44$ days) of the daylight hours foraging. In winter this is high compared with elk in the European Alps, Poland, Rocky Mountains, and Scotland (Clutton-Brock, Guinness, &

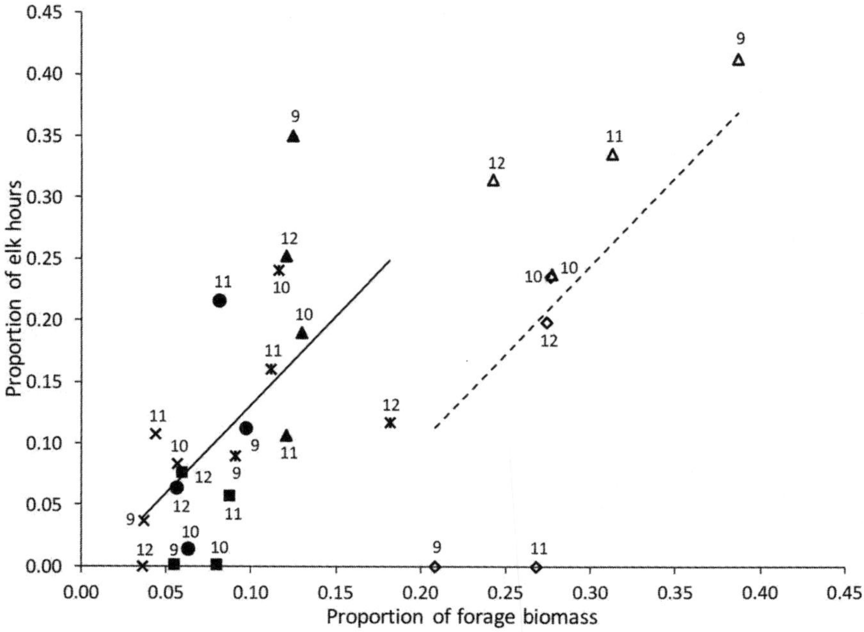

FIGURE 3.6. Scatterplot and regressions summarizing the relationship between forage biomass in a sector (proportion of the food supply in Davison Meadows) and elk forage use (proportion of elk foraging hours for the Davison herd that year) in that sector from 2009 to 2012. Regressions were when the Levee herd was present in the southern sectors (dashed line, sodav, C) and other sectors where the Levee herd was virtually absent (solid line). Numbers denote the year for that data point (e.g., 9–2009). Sector symbols were: open triangle—C, open diamond—sodav, closed diamond—A, x with vertical line—horsebarn, closed circle—B, closed square—picnic and x—WPC. Number of quarter-square meter plots in each sector were sodav—50, C—60, B—40, A—40, picnic—20, WPC—20, and horsebarn—40.

Albon, 1982; Georgii, 1981; Green & Bear, 1990; Kamler, Jedrzejewska, & Jedrzejewski, 2007). In those elk, no more than 50 percent of every 24 hrs was spent foraging. At night, the Davison herd appeared to be more sedentary and less active. The "not as strong as I would like" relationships between forage availability and use does not appear to be due to sedentary elk in the day that became more active and traveled farther at night.

FORAGE DEPRESSION OR SOCIAL FENCE?

The sodav and C sectors in the southern part of Davison Meadows was the partition zone, the area used by both the Levee and the Davison herds. Meadow partitioning might be because of grazing pressure by both the Davison and

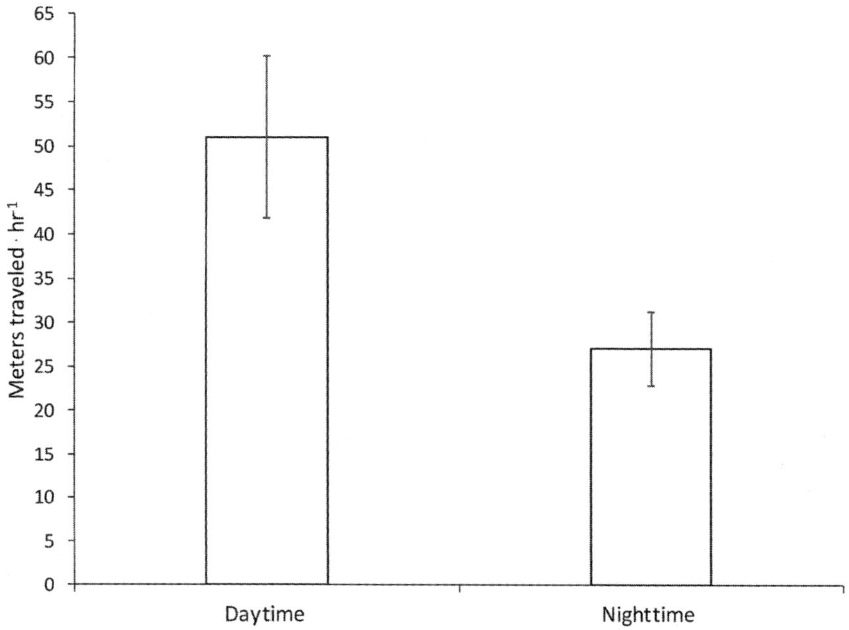

FIGURE 3.7. Bar chart of means, and standard error bars, of straight-line distance traveled per hour during daytime and nighttime hours for the Davison herd. Sample sizes were $n = 22$ in the day and at night. The Davison herd traveled farther in the day than at night (paired $t = -3.1$, $df = 11$, $P = 0.006$).

Levee herds. To explore this possibility, I compared forage biomass in the partition zone to the remainder of Davison Meadows between 2009 and 2012. I also compared the possibility of forage biomass decline across a 21-day period during my January field work in 2009, 2010, and 2011.

If there simply was less feed in the partition zone or grazing pressure was more intense, I would expect more forage biomass in the sectors used mostly by the Davison herd, the non-partition zone. The only trend that might be evident was that forage biomass (kilograms · hectares^{-1}) was greater, not less, in the partition zone (Fig. 3.8). The partition zone accounts for about half of the Davison Meadows, so it was not likely that the total forage biomass in the entire partition zone was less than in the remainder of the Davison Meadows. Perhaps the reason the Levee herd used the southern part of Davison Meadows was because of the forage abundance there.

Although the partition zone had as much forage as in the remainder of the Davison Meadows, grazing pressure in the partition zone might have resulted in greater depletion. Plant growth in January is probably not that much (Fig. 0.3). From one visit to the next, the Davison herd might have detected an

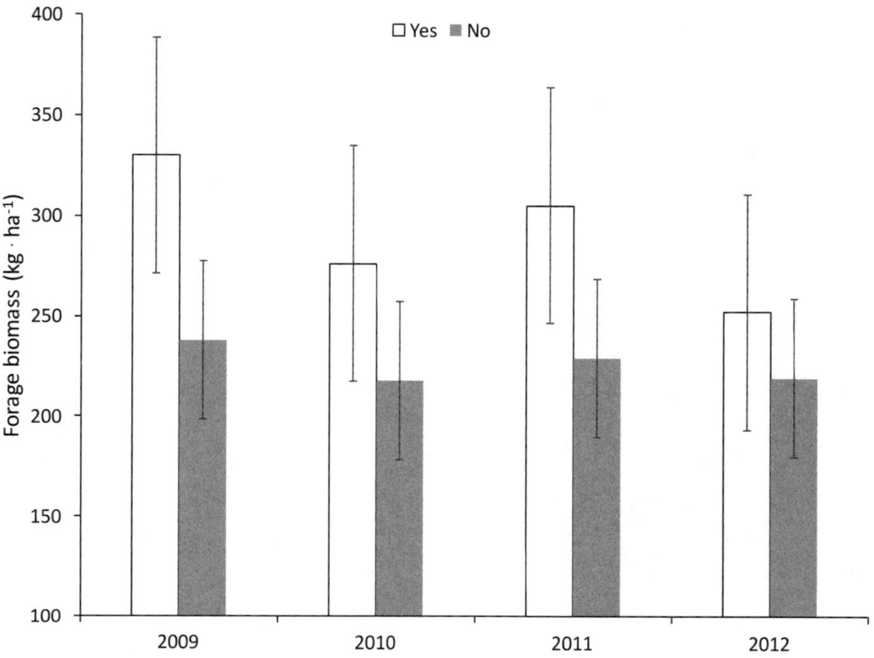

FIGURE 3.8. Forage biomass in sectors, and 95 percent CI, of the partition zone (sectors sodav and C, labeled yes) to the remaining sectors that were not in the partition zone (labeled no) 2009–2012. Forage biomass was measured at the end of the January field season. The linear mixed effects model had fixed factors of partition zone (yes, no), year, and a year X partition zone interaction. The interaction was weak ($F = 2.6$, $df = 3, 1067$, $P = 0.053$), yearly variation was evident ($F = 7.3$, $df = 3, 1067$, $P < 0.001$), and difference between partition zones was not evident ($F = 1.8$, $df = 1, 4.91$, $P = 0.233$). Furthermore, the model accounted for little variation in forage biomass. The conditional and marginal r^2 values were 0.07 and 0.22, respectively. Sector was the random factor.

accelerated decline of forage in the partition zone and chosen to forage in the partition zone less intensely. In 2009, I actually did detect a decline in forage biomass at the end of the January field season in the partition zone (Fig. 3.9). Nonetheless, in that year the decline was not statistically different from the decline in the non-partition zone. In the remaining 2 years, I found little to suggest forage depletion occurred in either the partition or non-partition zone. Furthermore, when data were pooled across years to increase sample sizes there also was little to indicate that forage depletion occurred in either zone. I found little evidence that would be consistent with the notion that the Davison herd avoided the partition zone because of a depleted food supply or just a smaller food supply.

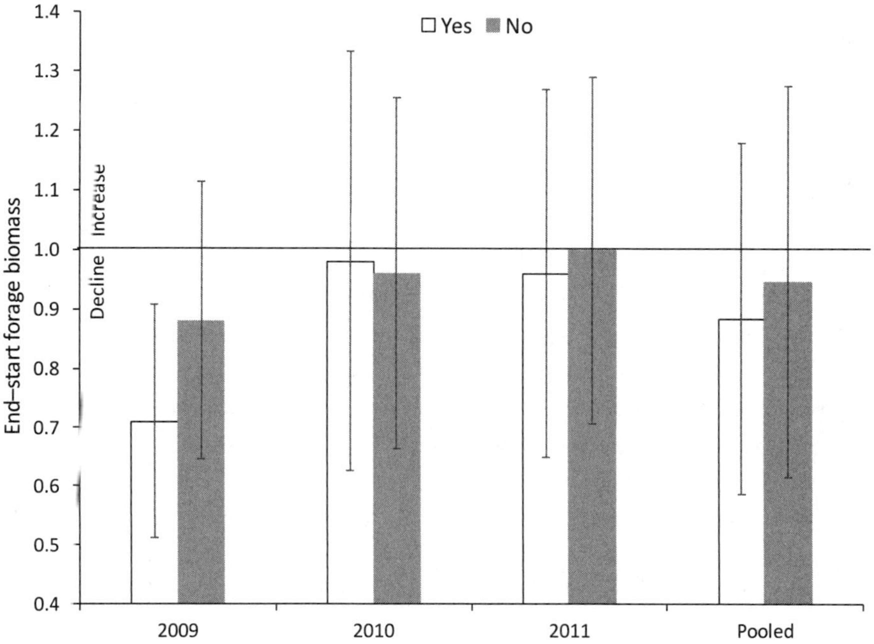

FIGURE 3.9. Difference between forage biomass, and 95 percent CI, in partition zone (yes) and remaining sectors (no) between 21 days at the end and start of the field seasons in January 2009, 2010, and 2011, and when years were pooled. Other factors were year and end–start. To estimate means I first analyzed a linear mixed effects model with all possible second- and third-order interactions as well as main effects. The analysis suggested complex patterns (partition zone × year × end–start interaction: $F = 4.3$, $df = 2, 1519.3$, $P = 0.014$). This model had transect and not sector as the random factor because the model with sector as a random factor had zero denominator degrees of freedom for the partition zone F ratio. Finally, I took the estimated means and variances from the mixed effects model to calculate end:start ratios and 95 percent CI according to Weckerly (1993) for each year and years pooled.

I need to say more about an effect on forage depletion from grazing pressure in the partition zone. My findings indicate directionality but no statistical significance. Indeed, forage depletion tended to be greater in the partition zone. There are numerous examples in ecology where individual studies addressing the same topic find no effect. Yet, when the findings across numerous studies are analyzed in one meta-analysis, an effect is detected (Gurevitch & Hedges, 1993). This is because the sample sizes of each study are insufficient to detect an effect. The sample sizes sufficient to detect an effect are only possible after analysis of findings from all studies. Because my study is the only one I am aware of that measured forage depletion by a large herbivore in a partition and non-partition zone, a meta-analysis is not possible.

Yet, my findings can be synthesized to talk about what should occur to obtain findings consistent with resource depression as the mechanism maintaining forage patch partitioning. In partition and non-partition zones, the key seems to be forage biomass in each of the zones and its depletion. If forage biomass is greater in a partition zone, its depletion should be greater than in a non-partition zone. Bite size and food intake by individual elk can then be greater than in the non-partition zone. If forage biomass in a partition zone declines below the biomass in a non-partition zone because of grazing pressure and the grazing pressure continues to keep forage biomass low, then resource depression is the likely explanation for partitioning of forage patches. Also consistent with resource depression as the mechanism would be low forage biomass in a partition zone because local environmental conditions do not favor abundant forage biomass. Clearly, this topic needs more study.

A social fence is a boundary between groups when there does not appear to be any discontinuities in forage habitat. The social fence hypothesis was developed by Hestbeck (1982). The central mechanism of Hestbeck's social fence is a positive feedback between population abundance and social impermeability of a group, in my case a herd. When food is limited, in other words at high population abundance, the costs of inter-herd aggression are greater than potential benefits from food acquisition. Herds that partition meadows are more likely to have a social barrier when food is more limited. Spatial distributions consistent with a social fence have been documented in elk (Mysterud et al., 2011).

One consequence of times when aggression is elevated should be more time with the head up when animals otherwise should be engaged in feeding. If an animal is subordinate, then he or she might be inclined to lessen opportunities for aggression by being on the watch for dominant or unfamiliar animals. In elk herds, juveniles are usually lowest in a social hierarchy (Franklin & Lieb, 1979). Greater vigilance should be most apparent in juveniles. Across all animals in a herd, the more time not foraging might pose real costs to food intake and serve as an impetus for animals to separate into their respective herds and go their separate ways away from the partition zone.

In January 2011 the Davison and Levee herds, for unknown reasons, decided to cooperate with me and fuse together in one aggregation for 3 days. I got lucky. This was the only time that I observed the 2 herds fused across the 4 years that I measured daytime activity of the Davison herd or, for that matter,

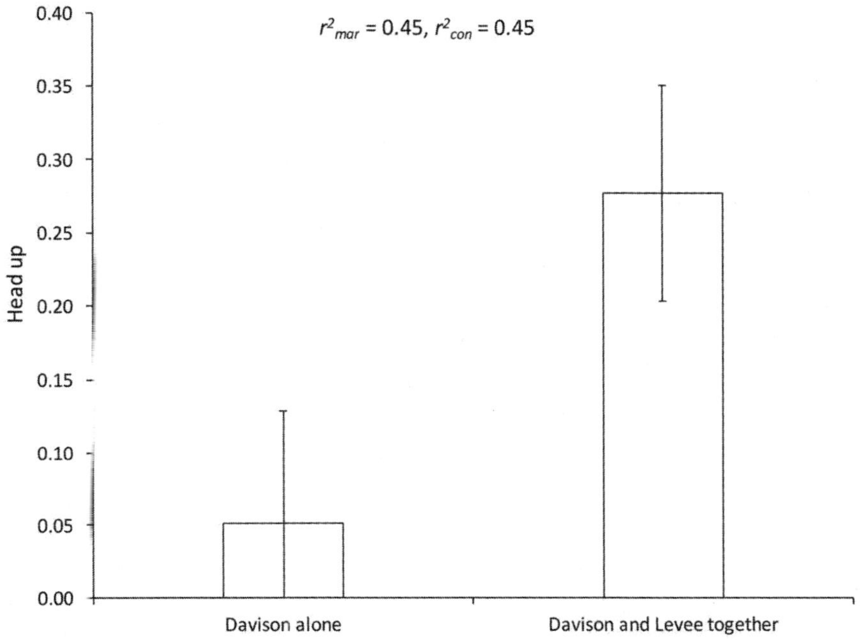

FIGURE 3.10. Estimated proportion of time, and 95 percent CI, that juveniles had the head up during focal observations collected when the Davison herd (n = 12) was alone (reference category) and when the Davison and Levee herds were together (n = 13) in one group in 2011. The linear mixed effects model had fixed factors of herd type (alone, together), forage biomass (kg · kg^{-1}), and length of the focal observation in minutes. The random factor was day that the focal observation was collected. The marginal and conditional r^2 values are also reported. The estimated regression of fixed factors (SE) were head up = 0.74(0.14) + 0.23(0.06) × herd type + 0.001(0.0004) × biomass + 0.062(0.02) × minutes. To estimate means, I used the average minutes (7), average biomass in sectors sodav and C in 2011 (220 kg · ha^{-1}), and group sizes of 34 and 55 for the alone and together group types, respectively.

across the entire 20 years of the study. Across those 3 days I measured time that the head was up in 13 juveniles. I measured the same thing in 12 juveniles from the Davison herd 2 days before the herds fused and 6 days after they split up and were separate herds again (Fig. 3.10). Juveniles had their heads up about 23 percent longer when the Davison and Levee herds were aggregated. When an elk had its head up for 5 percent of the time, I projected a food intake of 2.83 gm · min^{-1} versus 1.88 gm · min^{-1} when an elk had its head up for 28 percent of the time (using regressions in Figures 3.1 and 3.4, forage biomass = 220 kg · ha^{-1}, group sizes of 34 for Davison alone and 55 when Davison and Levee were together). Food intake by juveniles was projected to be 1.5 times greater when the Davison herd was alone. Social behavior might have ecological consequences that manifest into a social fence.

The partition zone appeared to be maintained by a social fence and not by forage depression in the partition zone. The Davison herd avoided the southern part of Davison Meadows, the partition zone, for social reasons. The question that needs some examination is this: Why did the Levee herd use the Davison Meadows at all? Across a time span greater than four years and considering all months of the year, there might have been interactions between herds that negatively affected herd sizes. As noted earlier, I did not detect the Levee herd in the Davison Meadows until 2007. From 1997 to 2002, I lived in the area. Across those years, I was in the RNSP much more frequently throughout the year than after 2002. Since 2003 my time in the RNSP has essentially been restricted to January. The Levee herd might have used Davison Meadows before 2007 and in months when I was not in the RNSP. This possibility seems particularly plausible after 2002.

Some insight might be gleaned by viewing the time series of counts of the two herds (Fig. 3.11). The herd counts seem, in general, to follow similar trajectories. Early in the time series counts were high; they declined in the middle years and increased thereafter. There was a positive correlation between counts of these two herds. If there were strong competitive interactions between herds that affected elk demography, however, I would anticipate an inverse correlation. The lack of an inverse correlation is consistent with no evidence of resource depression in the partition zone in Davison Meadows. Also, there appears to be no association between the herds in their intrinsic rate of population growths (lags of 1 or 2 years). The population dynamics of the Davison and Levee herds appear to be independent of one another. This finding concurs with Julian et al. (2013).

I suspect that the use of Davison Meadows by the Levee herd had something to do with what happened in the lower Redwood Creek drainage, an area just to the south of the Davison Meadows that the Levee herd had used throughout the entire time of my study (Fig. 1.5). In the lower Redwood Creek drainage, areas might have lost food and new food supplies might have suddenly become available. For example, in January 2015 and 2016, the Levee herd was not detected in Davison Meadows and yet the count of the Levee herd was the highest reported in the time series (59–60). The Levee herd fused with the nearby SOC herd (the herd that inhabited an area including the old South Operations Center of the RNSP) in autumn 2014, and those herds remained together thereafter. A cattle pasture suddenly became available to elk in autumn

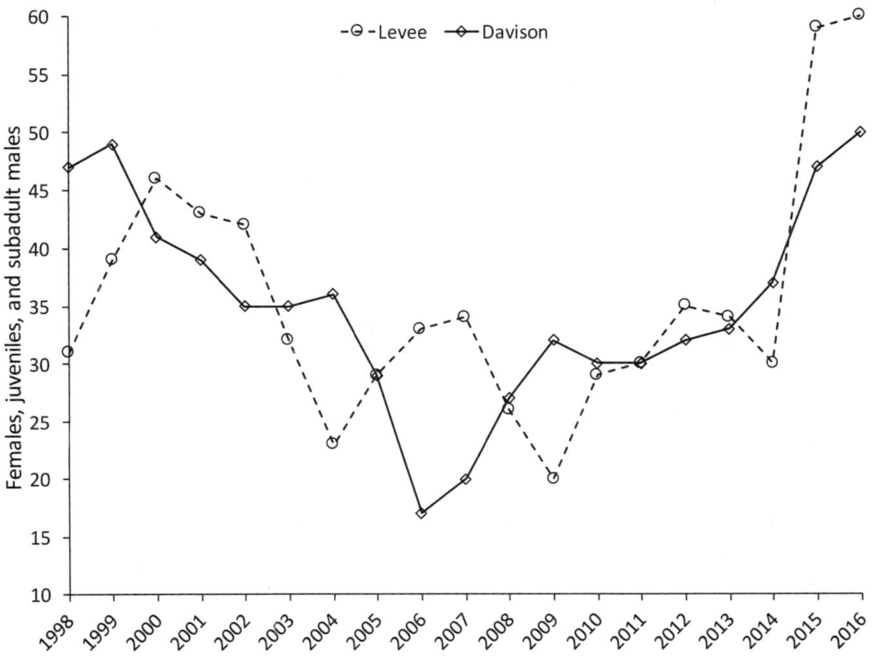

FIGURE 3.11. Time series of abundances of Davison and Levee herds from 1998 to 2016. The correlation between herd abundances was $r = 0.87$ ($df = 17$, $P = 0.01$). Yet, there was no correlation between intrinsic rates of population growth between the two herds with lags of one (ln(count in year $t + 1$/count in year t), $r = -0.16$, $df = 16$, $P = 0.524$) or 2 years (ln(count in year $t + 2$/count in year t)/2, $r = -0.18$, $df = 15$, $P = 0.483$).

2014, on the south end of the town of Orick. Before that time, this pasture had been fenced off and unavailable to elk (Hofstra, 1997). For reasons unknown to me the gates were opened, allowing elk access. The two herds in the vicinity, Levee and SOC, began using the cattle pasture and fused. The Levee herd might not have visited the Davison Meadows in 2015 and 2016 because it was exploiting a new food supply.

The manifestation of Hestbeck's social fence relies on a positive feedback between population abundance and social impermeability. As population abundance increases, food becomes more limited and the costs of aggression should be great in relation to benefits to foraging. What seems at odds with a positive feedback is the abundance data of the herds from 2009 to 2012 when the social fence data were collected. The abundance of the Davison herd was intermediate from 2009 to 2012 (Fig. 3.11). The size of the Levee herd was more variable. In 2009 the size of that herd bottomed out at 20 and then increased

1.5 times to 35 by 2012. In early winter, food supplies were low and spatial overlap between herds was low compared to other seasons of the year (Kolbe & Weckerly, 2015). At that time of year, a social fence might be able to manifest across a range of population abundances.

INTERPRETATION

The distribution of herds in the lower Redwood Creek and Prairie Creek drainage appears to be influenced by the distribution of open habitat which, during the time of this study, was primarily meadows. Meadows can be close to one another or they can be separated by expanses of forest that do not contain nearly as much forage for elk as do meadows, particularly in winter. If elk in a herd can ingest sufficient amounts of forage from a meadow to survive, grow, and reproduce, and the meadows are situated near other meadows, then one or more herds might partition these meadows. This appears to be the case with the Davison Meadows. Meadow partitioning should foster coexistence between herds by reducing resource competition. Partitioning of the Davison Meadows did not seem to be driven by resource depression but by a social fence. At least the Davison herd appeared to avoid the area frequented by the Levee herd. The reason for avoidance seemed to be coupled to aggression among elk unfamiliar with one another.

Herds are comprised of socially bonded females, juveniles, and subadult males, animals that are together most if not all of the time. Social bonding is probably associated with landscape attributes such as discrete patches of forage habitat that are dispersed in other habitat that has much less forage. Furthermore, the forage habitat patches are not ephemeral but long lasting. In comparison, social bonding between elk appears to be lower in landscapes where open habitat is more expansive (Franklin & Lieb, 1979). Individuals inhabiting expansive, open landscapes seem to aggregate with differing sets of individuals regularly. Discrete food patches that are long lasting and embedded in an expanse of habitats with little forage seem to facilitate the establishment of bonds between elk that are long lasting (Street et al., 2013; Weckerly, 1999). Young animals might benefit from older animals by learning about the distribution and abundance of forage that change across seasons as well as about lethal risks in the landscape (Sueur et al., 2011).

Elk in the Davison herd moved as one and, most likely, used the heterogeneous food supply in Davison Meadows efficiently. Areas were used in relation to forage biomass. The Davison herd foraged in most or all sectors in the month of January. Usually they foraged in sectors on multiple occasions within the month. The herd certainly did not move around Davison Meadows at random. If that had been the case I would not have detected a relationship between forage biomass and hours that elk in the herd foraged. Using food supplies in relation to availability seems to update memory and maintains an animal's spatial familiarity with food distribution and abundance. Spatial familiarity with the environment allows animals to effectively exploit a predictable food source. An outcome of updating and maintaining knowledge about a food supply that is spatially and temporally predictable is using a small area when the regrowth rate of grazed forage is high (Fryxell et al., 2005; Merkle et al., 2014; Riotte-Lambert et al., 2015). Indeed, the 51 ha of the Davison Meadows is a small area for Roosevelt elk to use on a monthly basis (Franklin et al., 1975; Kolbe & Weckerly, 2015).

Although herds rarely join together as one, when it occurred it was revealing in terms of how a social fence might be maintained. First, herds might fuse together in a partition zone because food is more plentiful. In January, elk using areas with more forage should see noticeable increases in food intake because of dramatic changes in bite sizes associated with small changes in forage biomass when that biomass is no more than 500 $kg \cdot ha^{-1}$. The benefits to food intake in the partition zone, however, might not be realized by all herd members. When herds are joined together as one, subordinate animals are probably spending more forage time than usual being vigilant or watching out for dominant and, perhaps, unfamiliar animals. An elk with a head up cannot ingest food because it probably does not multitask (Fortin, Boyce, Merrill, & Fryxell, 2004). Multitasking is when animals can ingest food and watch out for potential risks at the same time. To understand multitasking, imagine summer in the RNSP. Food is more abundant and elk bite sizes should also be large. Animals in that season that are watching out for something might also have the opportunity to chew the food that is already in their mouths. Bite sizes in winter should be considerably smaller than in summer. An elk that is watching out for something in winter probably does not have a lot of large food particles that require chewing. The time the head is up is potentially costlier in winter than in summer.

Elk that do not graze cannot ingest forage, an intuition that seemingly requires no remark or study. The foraging process in large herbivores, however, is complex. Understanding the nuances of foraging helped me determine how a social fence could manifest in elk. Of course, my evidence for a social fence is based on data gathered from one partition zone. Although it was a somewhat detailed examination, the sample size is measly—one partition zone. Whether my conclusions are applicable to other foraging patches partitioned by herds is unknown.

The findings about the living space or distribution of herds suggest that inter-herd comingling was rare. As such the abundance dynamics among herds might differ as well. Indeed, the findings in the next three chapters reveal that abundance dynamics differed among herds.

4

The Unexpected

Population Irruption

Whenever an ungulate population is faced with a standing crop of vegetation in excess of that needed for maintenance and replacement of animals, an irruption and crash is the inevitable consequence.

— Caughley, "Wildlife Management and the Dynamics of Ungulate Populations"

Professor Aldo Leopold (1943) was the first to describe a complex pattern of population dynamics displayed by four deer (*Odocoileus*) populations: one was on the Kaibab plateau, Arizona, and the other three were in the upper Midwest. He labeled the complex pattern "irruptive dynamics." Perhaps because of the complexity, irruptive dynamics were defined by four stages (Fig. 4.1).

The first stage is rapid population growth that is seemingly unchecked by limitations in the food supply and from being a food supply to bears, puma, wolves, or any other predators. The ungulate food supply during the rapid-growth stage can be described as a surplus (Caughley, 1976; Forsyth & Caley, 2006).

The second stage is the overshoot. The rate of population growth is so rapid during the growth phase that the population literally overshoots an exhausted food supply. Forage plants do not have the time to adjust to the intense herbivory.

The third stage is population decline, precipitated by population overshoot. As a consequence of the depleted food supply, the population is reduced, either slowly or abruptly, to an abundance noticeably lower than during the overshoot. Usually during the decline stage there are changes to the ecosystem set in motion by intense herbivory during the rapid growth and overshoot stages. Ungulate herbivory changes plant communities and indirectly

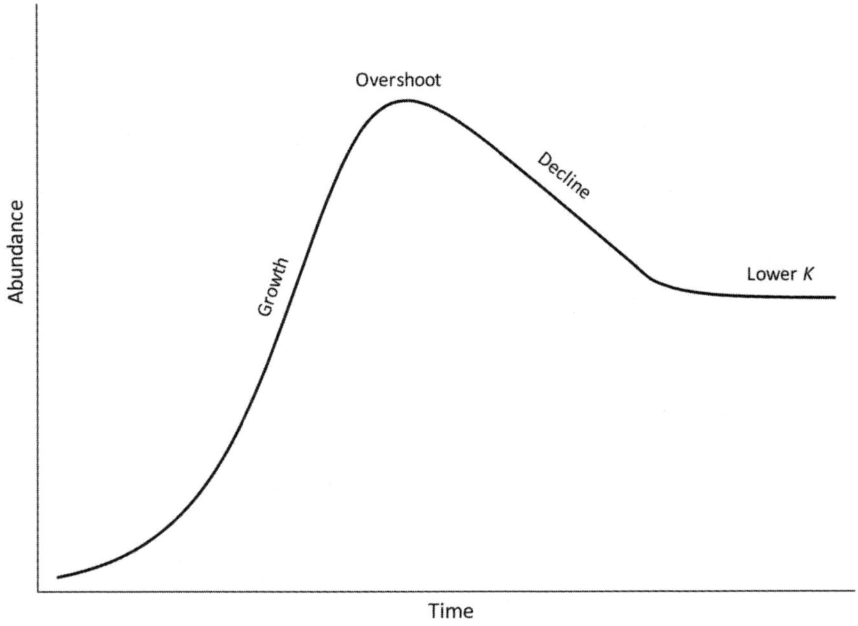

FIGURE 4.1. Conceptual diagram of a population irruption with each of the four stages labeled on the graph.

ungulate grazing can affect nutrient cycling (Ricca, 2013). At some point the decline stops.

Finally, the fourth stage of the irruption is the ungulate population then arriving at a new equilibrium with a changed food supply at a lower *K* carrying capacity. The lower *K* is in reference to the abundance and presumed initial *K* at the peak of the overshoot stage.

Since Leopold (1943), ungulate irruptions have been documented in numerous populations with a number of ungulate species (Ricca, 2013). Sometimes the irruptive label can be applied loosely. A population might display rapid and exponential growth so the irruptive label is applied. Yet, there is no follow-up or reporting of what occurs after rapid growth. I reserve the irruptive label for populations that display all four stages and obtain a new lower *K* carrying capacity. A population might display rapid growth but then remain abundant (Cobb, 2010; Le Saout et al., 2014; McCullough, 1997; Simard, Cote, Weladji, & Huot, 2008). Frequently documented examples of irruptive dynamics occur on oceanic islands at high latitudes (Tyler, 2010). The irruptions follow a similar pattern: A large herbivore is introduced by people to an island that was not previously inhabited by the ungulate or any ungulate species in recorded history. The herbivore then has access to a food supply that has little tolerance or resistance to being fed on by ungulates. A tolerant plant can grow and reproduce even after it has been defoliated by ungulate consumption. A plant resistant to consumption by a large herbivore possesses chemical compounds or physical structures that render it unpalatable; as a result, the ungulate does not eat it or finds it difficult to consume or digest. The island also has no natural predators to reduce the rate of rapid growth. Finally, ungulate die-offs initiating the decline stage have been associated with interactions between density-dependent and density-independent factors. Animals are in poor physiological condition because of numerous mouths consuming a limited food supply. Thus, animals are more vulnerable to succumbing to climatic factors such as deep snow and ice (Klein, 1968).

The species that have been "liberated" on high-latitude oceanic islands are usually caribou or reindeer (*Rangifer tarandus*). Once released, the animals consume a food supply at the introduction site that does not renew itself. A primary winter food, reindeer lichen (*Cladonia*), is intolerant to caribou herbivory and does not return for years, or maybe decades, after decimation from grazing (Klein, 1987; Ricca, 2013). The animals then immigrate to new areas on the island to forage. The immigration can be organized along a dispersal front (Caughley, 1970; Ricca, Van Vuren, Weckerly, Williams, & Miles, 2014). As long as there are new foraging areas to exploit, the dispersal front continues to move into ungrazed land and the population continues to grow rapidly. Once the dispersal front arrives at the beach, there are no new places to exploit forage. The conditions are right for an overshoot and decline, followed by a new *K* carrying capacity that is lower than the abundance at the overshoot.

Population dynamics of elk at Boyes Meadows from the 1930s to the 1960s suggest an irruption (Fig. 1.3). But I could not imagine the possibility of one for much of the 20 years of my study. Although meadows in the forested landscape of the RNSP can be viewed as forage habitat islands, there was little else about the environment that seemed to conjure a setting conducive to an irruption (Starns et al., 2015). Despite the insular pattern to forage habitat there were no physical barriers to prevent movement among meadows. Davison Meadows and Boyes Meadows were also rather small. I have observed elk walk across each of these meadows in under an hour. The establishment of a dispersal front in these meadows on the time scale of months or years is unrealistic. Abundances of bears and puma have not been estimated in the RNSP, but in all likelihood these predators have been ubiquitous across the time of my study and perhaps since the 1980s. The mild climate seems to exert little density-independent limitations on the Davison herd (Starns et al., 2014). Furthermore, what weighed most heavily on my mind was that elk were a part of the landscape for at least 1,000 years. Plant communities presumably have experienced elk feeding for a considerable period of time. Within the past 150 years, if elk grazing did not occur, then such grazing was replaced by domestic livestock grazing. The only surplus food supply I could imagine was a barn filled with hay that elk could consume at their want. Of course, no such barn existed.

Yet, I am calling the Davison population dynamics an irruption (Starns et al., 2014; Starns et al., 2015). Before 1991 Davison Meadows was privately owned and grazed by cattle (Richey & Wheeler, 2001). Davison Meadows apparently was forested before 1890. When the Davison family purchased the land in 1890 they began clearing forest, which became the Davison Meadows. For much of the time between 1890 and 1991 cattle grazed the meadow. Details of cattle grazing, such as the numbers, which meadows were grazed, or the seasonal and annual occurrence of grazing, is unknown. Elk, however, did seek to graze Davison Meadows before 1991. In the mid-1980s elk depredation permits were issued and 15 elk were taken from the lower Redwood Creek drainage and the Davison Meadows (Hofstra, 1997). How many of the 15 elk were removed from Davison Meadows is unknown. In 1991, when Redwood National Park took ownership, cattle grazing ceased and elk had unhindered access to Davison Meadows.

Although elk probably grazed Davison Meadows before 1991, I consider that year as the year of elk colonization because elk then had unimpeded access to that area (Fig. 4.2). When I began population surveys in 1997, the Davison herd (females, juveniles, and subadult males) and the total population (all age–sex classes) were the largest I had documented. Because the influence on the food supply in Davison Meadows would be from all elk, I report total population size as well as Davison herd abundance. The irruption is most evident in the dynamics of the total population. After 1997 a steady decline occurred until 2006 when abundance bottomed at 32 elk. After the nadir, the population increased until 2009. Between 2009 and 2012 the total population size fluctuated between 36 and 44 animals. Since 2013 the population has increased to the 51 elk counted in 2016. The total population size in 2016 was roughly 61 percent of the size it was in 1997. What is striking is that adult males (total population size–Davison herd abundance) composed a larger part of the total population early and not later in the time series. Consequently, if you only eyeball herd abundance, the dynamics appear cyclical more than irruptive. Nonetheless, since 1991 the four stages of the irruption can be more clearly delineated from

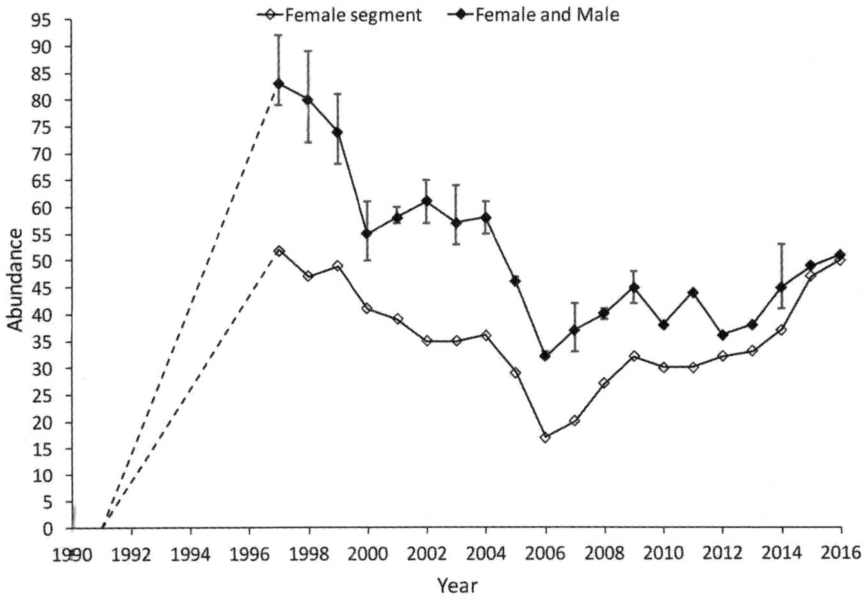

FIGURE 4.2. Abundance estimates of females, juveniles, and subadult males (female segment) and total population (female segment and male) in the Davison Meadows, 1991–2016. The Davison Meadows were purchased by the National Park Service in 1991 and became available to elk that year. My systematic population surveys began in 1997 so the dashed lines were years when no formal population surveys were conducted. In some years, every adult male sighted in Davison Meadows was uniquely identified. Abundance in those years was the sum of high count of the female segment and the number of marked adult males. I used Bowden's mark-resight estimator to calculate total abundance in years when there were unmarked adult males sighted in Davison Meadows during surveys (Weckerly, 2007). The error bars are 95 percent CI. The nadirs or lowest abundances were in 2006.

total population size: rapid growth between 1991 and 1997, the peak or overshoot in 1997, decline between 1998 and 2006, and lower *K* carrying capacity thereafter.

I considered the setting of the Davison irruption unusual. Consequently, I sought to determine the factors precipitating the rapid growth stage and whether density dependence changed between the decline and lower *K* stages. I also wanted to estimate the lower *K* carrying capacity and assess the surplus food supply. Irruptions are set in motion by conditions that allow the colonizing population to grow rapidly. The decline stage is often a population crash because of a die-off, implying that there are interactions between density-dependent and density-independent factors (Tyler, 2010). Yet, in the mild climate of north-coastal California, density-independent factors appear to have little influence (Starns et al., 2014). Irruption population dynamics governed solely by density-dependent processes have not been explored in much detail.

Estimating *K* carrying capacity of herds in the RNSP is needed to assess resilience, one of the goals of this book. Finally, there probably were plants prevalent before the decline stage that became scarce and were replaced by other species that were resistant to elk grazing by the lower *K* carrying capacity stage (Starns et al., 2015). Those plants, whatever they might be, would be considered the surplus food supply.

RAPID GROWTH

The growth of the total population and Davison herd from 1991 to 1997 was indeed rapid. Calculating the intrinsic rate of maximum increase, $r_{max} = \ln(N_{1997}/N_{1991})/6$, where *N* signified abundance, r_{max} was likely quite high. For example, if there were two elk (one adult male and one adult female) that colonized Davison Meadows in 1991, r_{max} for the Davison herd would be 0.543 and 0.621 for the total population. In other populations of elk colonizing habitat with abundant food supplies, few predators, and where rapid growth appeared to be driven solely by demography and not movement, r_{max} ranged from 0.23 to 0.30 (Eberhardt, Eberhardt, Tiller, & Cadwell, 1996; Komers & Curman, 2000; McCorquodale, Eberhardt, & Eberhardt, 1988; Sargeant & Oehler, 2007). It defies explanation to start with two elk and have the abundance observed in 1997 entirely from recruitment and high female survival. Immigration had a part to play in rapid growth.

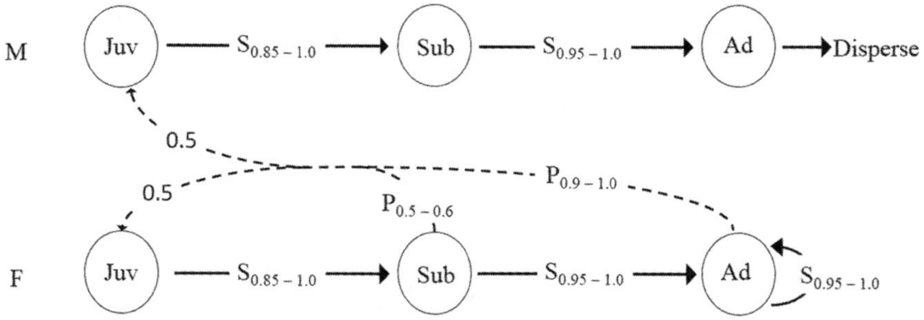

FIGURE 4.3. Schematic of vital rates for exponential population growth between June 1, 1991, and June 1, 1996, in the Davison Meadows. I assume that abundance on June 1, 1996, will be similar to abundance the following January 1997. In 1991 there were either 8, 10, or 12 sexually mature females that colonized the meadows. These females were at least six years from senescence. The notation is M–males, F–females, juv–0 to 1 year of age, sub–1 to 2 years of age, ad–> 2.0 years of age, S–survival rate, and P–pregnancy rate. The range of values in subscripts are the values that could be randomly drawn from a uniform distribution. The sex ratio at birth was 1:1. There were 10,000 iterations of the simulation and the median, 2.5th and 97.5th percentiles (CI) were reported. Ranges for vital rates were gleaned from Eberhardt et al. (1996), Murphy (1963), and Sargeant and Oehler (2007).

To gauge the role of immigration to rapid growth, I conducted a simulation projecting growth of the Davison herd from June 1, 1991, to June 1, 1996 (Fig. 4.3). Immigration is when an elk that did not reside in Davison Meadows travels to Davison Meadows and then establishes a home range that includes Davison Meadows. I focused on the herd (i.e., females, juveniles, and subadult males) because these animals can provide population growth through movement and juvenile recruitment. Because of the rapid population growth between 1991 and 1997, I assumed animals were on a high nutritional plane and in the best possible condition to achieve high vital rates and exponential growth. I compared the herd abundance and the juvenile:female ratio (juvenile recruitment) projected on June 1, 1996 to the corresponding statistic calculated from my population surveys in January 1997. I assumed all births had occurred by June 1. Projected abundance on June 1, 1996 might be a little higher than my observed data collected seven months later because some elk could have died. Juvenile:female ratios might also be influenced.

The findings from the simulation indicated that 10 adult females colonized Davison Meadows during the rapid growth stage (Fig. 4.4). The projected abundance of the Davison herd with 10 colonizing females was 51 (95 percent CI: 46–56) with a juvenile:female ratio of 0.59 (0.54–0.65) in 1997. This is what I anticipated under exponential growth. The juvenile:female ratio was calculated using the sum of subadult and adult females as the denominator. In environmental settings with an abundant, nutritious food supply, 50 to 60 percent of subadult females can conceive and give birth to one young each (Sargeant & Oehler, 2007). Adult females also have a litter size of 1.0 but a higher pregnancy rate (0.9–1.0). Under conditions favoring exponential growth, a higher pregnancy rate would suggest a juvenile:female ratio closer to 1.0 for adult females. The reason the projected ratio was 0.59 was that subadult females had lower fecundity. Nevertheless, the projected juvenile:female ratio is roughly double the observed ratio in 1997 (0.30). It seemed unlikely to me that the lower observed ratio could be solely because of deaths between June 1996 and January 1997. The lower observed juvenile:female ratio indicates that female fecundity and juvenile survival might have begun to decline before 1997 because animals were on a lower nutritional plane with reduced protein and fat stores (McCullough, 1979). Assuming exponential growth seemed to have dramatically overestimated observed juvenile recruitment in 1997, which means that actual female immigration was probably underestimated.

I am not confident that the simulation precisely estimated the number of females that immigrated into Davison Meadows. But, the simulation did show that ingress of animals had a substantial role during the rapid growth stage of

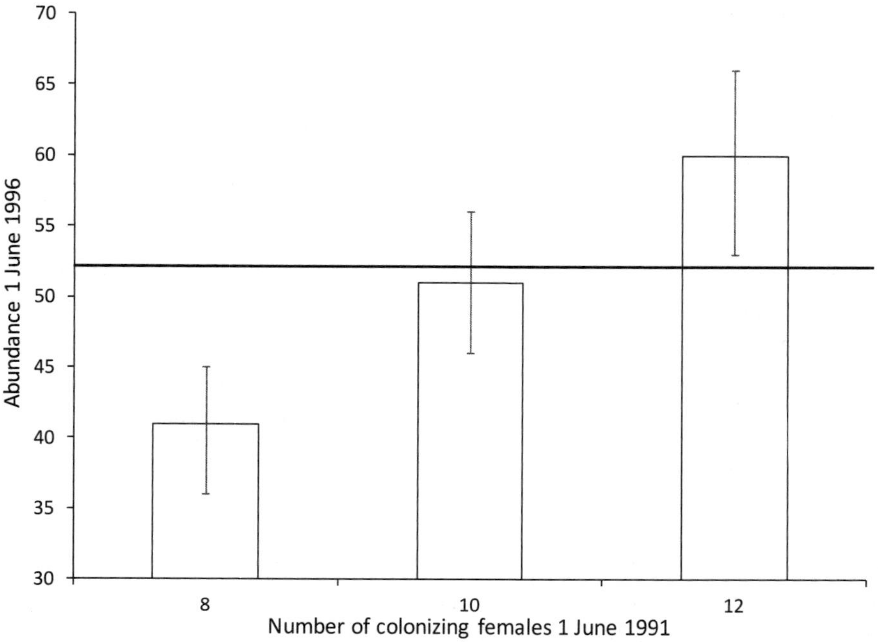

FIGURE 4.4. The simulated mean abundance of females, juveniles, and subadult males, and 95 percent CI, on June 1, 1996, when 8, 10, or 12 adult females presumably colonized Davison Meadows on June 1, 1991. Each simulation had 10,000 iterations. The horizontal open bar is the number of females, juveniles, and subadult males observed in January 1997 (52). The findings from the simulation also indicated a juvenile:female ratio of 0.62 (95 percent CI: 0.55–0.67) on June 1, 1996, when 10 adult females presumably colonized Davison Meadows.

the irruption. I did observe one likely immigration by a female to the Davison Meadows. In the 1970s, when the land was privately owned, logging occurred to the southwest of Davison Meadows. In the mid-1980s Humboldt State University was capturing and ear-tagging elk as part of a radio telemetry study to examine spatial patterns (Hofstra, 1997). One of the capture areas was the open forests to the southwest of Davison Meadows. One of those ear-tagged females captured in 1986 was a member of the Davison herd in 1997 (Richard Golightly, personal communication).

If immigration was primarily by animals occupying forest and open forest habitat, then it might have been driven by ecological rewards and risks associated with the landscape composition of habitats. On a per-hectare basis, forests and open forests have markedly less elk forage than meadows (Weckerly, 2005). By the early 1990s, most of the forest opened in the 1970s probably had stands of conifer and redwood trees that were closed canopied or approaching that state. When elk were given unhindered access to Davison Meadows, animals in forested habitat might have sought out Davison Meadows. Whether elk remained in Davison Meadows when elk first encountered that meadow was probably dictated by the food supply in Davison Meadows relative to the food supply in the surrounding forest. Risks posed by natural predators also should be considered. If the per capita food supply was greater in the meadows, and elk also were less vulnerable to predation, then elk should use meadows more than forests (Heard, 1992). Add to the mix the relatively small size of the Davison Meadows (51 ha). The size of Davison Meadows precluded the development of a dispersal front as found on oceanic islands that are substantially larger (Ricca et al., 2014). Forage in all parts of Davison Meadows was probably exploited shortly after those meadows became available to elk. As such, during the rapid growth stage immigrating elk likely continued to use Davison Meadows. The resulting nutritional plane restricted female fecundity and juvenile survival to a point where rapid population growth could not have been fueled entirely from juvenile recruitment and high adult survival rates.

ADULT MALES

Again, what strikes me about the dynamics of the Davison population is the abundance of adult males (total abundance–herd abundance), relative to herd abundance, at the overshoot stage in 1997 and then the subsequent decline through the remaining time series (Fig. 4.2). The decline in the numbers of adult males does not seem to be from a decline in recruitment of subadult males by Davison females that might remain in the area (Fig. 4.5). Across 1997 to 2016 the annual proportion of the Davison herd comprised of subadult males was similar.

A decline in the number of adult males, relative to females, as a population approaches *K* carrying capacity has been observed in other elk populations that were not hunted (Cobb, 2010; Coulson, Guiness, Pemberton, & Clutton-Brock, 2004). The decline in males with increasing population abundance might be associated with males seeking mating opportunities when both males and females display a high degree of overlap in use of resources. In this instance, the resource is Davison Meadows. Males often disperse from the natal herd around two years of age or at the conclusion of the subadult life stage (Franklin & Lieb, 1979; McCullough, 1969). The tendency to disperse and the dispersal distance seems to be dictated by both resource competition and potential

FIGURE 4.5. Line chart of annual abundance of number of subadult males expressed as a proportion of the number of females the previous year (lag 1), 1998–2016. I assumed all spikes counted during surveys were recruited from the Davison herd. There was no correlation between year and proportion of subadult males ($r = 0.15$, $df = 16$, $P = 0.885$).

mating opportunities (Loe et al., 2009). As subadults, males are still growing and should be prone to disperse if forage is limited (Fig. 2.1). Yet, if forage in an area is not overly limited and there are numerous females, then males might be inclined to settle in such areas regardless of whether the area is near to where they were born or tens of kilometers away (Loe et al., 2009). Of course, if males decide to remain near to their natal area, then there should be a greater possibility of inbreeding. Nevertheless, the stage of irruption might complicate male decision making. During the rapid-growth stage, males should be attracted to Davison Meadows and might remain until density-dependent feedbacks between elk and the food supply begin to emerge. Food should definitely become more limited during the lower *K* carrying capacity stage. Thus, there should be an inverse relationship between abundances of the Davison herd and adult males by the lower *K* carrying capacity stage.

Females should benefit from high use of Davison Meadows, with its rich food source relative to forest, even when food is more limited at *K* carrying capacity (Clutton-Brock et al., 1982; Kie & Bowyer, 1999; McCullough, 1999). At most, females can produce one young per year. Whether that happens depends on protein and fat stores (Cook et al., 2013). Also to consider is that female demography is tied to herd abundance and not to total abundance. Female and male ungulates dimorphic in body size display resource partitioning (Clutton-Brock et al., 1982; McCullough, 1979). Herd abundance should be more strongly related to itself than to male abundance. Finally, the irruption itself has to be factored into the mix. Relationships might differ during the decline and lower *K* carrying capacity stages (Starns et al., 2014).

To assess these possibilities, I conducted model selection analyses. I estimated log-linear or Poisson models (regressions) that had abundance in year *t* as the response variable and abundance in the previous year (lag one) as a predictor. If the lag one coefficient was a negative value and statistically significant, then it indicated direct density dependence (Dennis et al., 2006). Abundance in one year was coupled to abundance in the next year because the number of mouths consuming a finite food supply influences demography. I evaluated two possibilities:

1. Whether abundances of adult males and the Davison herd were influenced by the abundance of males or the herd lagged 1 year.
2. Whether relationships differed between the decline and lower *K* carrying capacity stages or whether the break in the relationship was in some other year. As a refresher, the abundance nadir in 2006 delineated the decline from the lower *K* stages.

Adult male abundance was influenced by the Davison herd abundance lagged 1 year, not adult male abundance, and the relationship differed between 1998–2008 and 2009–2016 (Table 4.1). There was some indication of a positive relationship between 1998 and 2008 but thereafter the relationship was inverse. Davison herd abundance, in contrast, was positively influenced by its own abundance the previous year, and not by adult male abundance, and the relationship did not change between the two irruptive stages. For the Davison herd analysis, there were two other models with deltas (Δ) less than 2.0. Yet, both of those models had two additional parameter estimates, and the coefficients estimating those two additional parameters were not influential ($P > 0.46$).

Neither adult male nor the Davison herd abundances seemed to be regulated by abundances before 2008. More females, juveniles, and subadult males the previous year seemed to beget more adult males as well as females, juveniles, and subadult males the next year. In the early years of the study, herd

TABLE 4.1. Summaries of AIC_C analyses of log-linear, first order autoregressive models (year *t*–1 vs. year *t*) estimating influences on abundance of adult males and the Davison herd (females, juveniles, and subadult males) between 1998 and 2016. Same-sex (SS) refers to adult males or the Davison herd and vice versa for opposite-sex (OS). The log-likelihood is abbreviated LL, number of parameters is nPar, and Δ is the difference between AIC_C of a model and the model with the smallest AIC_C. Nadir was a categorical predictor representing years including and before the smallest population size in 2006 (reference category) and years afterward. Categorical break variables allowed for relationships to differ in intercepts and slopes between periods of years besides the nadir. For example, break4 estimated separate regressions for data between 1998–2004 and 2005–2016. The model selected for each response variable is marked with an asterisk. Below are the parameter estimates for each selected model. Because the assumption of independence might be violated, parameter estimates and standard errors were obtained from bootstrapping. The X denotes an interaction.

		ADULT MALE		DAVISON HERD	
MODEL PREDICTORS	NPAR	LL	Δ	LL	Δ
Lag 1 abundance same-sex (SS1)	2	−55.16	11.31	*−58.39	*0.00
SS1, break4, break4 X SS1	4	−54.82	10.64	−57.71	4.75
SS1, break5, break5 X SS1	4	−54.76	10.51	−57.35	4.01
SS1, nadir, nadir X SS1	4	−55.16	11.31	−55.83	0.98
SS1, break7, break7 X SS1	4	−54.08	9.15	−55.72	0.77
SS1, break8, break8 X SS1	4	−54.79	10.58	−56.37	2.05
Lag 1 abundance opposite-sex (OS1)	2	−92.03	78.96	−72.73	28.67
OS1, break4, break4 X OS1	4	−60.22	21.44	−58.57	6.46
OS1, break5, break5 X OS1	4	−53.64	8.26	−57.94	5.21
OS1, nadir, nadir X OS1	4	−51.49	3.97	−61.09	11.51
OS1, break7, break7 X OS1	4	−51.37	3.73	−61.06	11.45
OS1, break8, break8 X OS1	4	*−49.50	*0.00	−61.09	11.51

	ADULT MALE				DAVISON HERD			
COEFFICIENT	ESTIMATE	SE	*z*	*P*	ESTIMATE	SE	*z*	*P*
Intercept	2.410	0.292	8.24	< 0.001	2.713	0.144	18.84	< 0.001
OS1	0.016	0.008	1.87	0.061				
Break8	4.536	1.826	2.48	0.013				
Break8 X OS1	−0.177	0.055	−3.20	0.001				
SS1					0.023	0.004	6.38	< 0.001

abundance was high and males were numerous. Later, in the decline stage, herd abundance declined as did abundance of adult males. After 2008 adult male abundance was inversely related to female, juvenile, and subadult male abundance the previous year. This finding suggests that forage was more limiting to adult males.

It is also possible that after 2008 adult male abundance in the Davison Meadows might have been influenced by what occurred in Boyes Meadows 3 km to the north (next chapter). By 2008 the abundance of the Boyes herd was low and after 2010 I failed to encounter any females in Boyes Meadows during

January surveys. Between 2011 and 2016 forage habitat near the Davison Meadows was seemingly vacant of females, juveniles, and subadult females, including the Davison herd. After 2008, adult males might have been inclined to use Boyes Meadows and not Davison Meadows.

Abundance of the Davison herd was affected by herd and not by male abundance. Females appear to partition resources from males similar to what has been reported in other populations of dimorphic ungulates (Clutton-Brock et al., 1982; Kie & Bowyer, 1999; McCullough, 1979). An intriguing pattern was that the Davison herd had a positive, and not a negative, relationship between abundances of the previous and current year. In order for *K* carrying capacity to develop, abundance has to be affected by density-dependent feedbacks between resources and demography that influences abundance. This pattern suggests that Davison herd abundance has yet to come into equilibrium with a reduced food supply or arrive at a lower *K* carrying capacity. Whether the Davison herd abundance is approaching a lower *K* carrying capacity requires further exploration, which is presented in the next section.

DENSITY-DEPENDENT RECRUITMENT

Irruptions often occur with die-offs of animals that are associated with climatic factors (Tyler, 2010). The die-offs usually delineate the initiation of the decline stage. Perhaps the most dramatic die-off was the population crash on St. Matthews Island, Alaska, when over one winter with much snow and ice the reindeer population dropped from a lot (6,000) to very few (42) animals (Klein, 1968). No such abrupt die-off was apparent in the Davison irruption (Fig. 4.2). The decline here was gradual.

If the mild climate of north-coastal California had density-independent influences on elk demography, it probably would manifest through juvenile recruitment. In large ungulates such as elk, adult survival is often high and does not display nearly as much variation as recruitment of juveniles (Gaillard et al., 1998). For reasons associated with large size, adults are less likely to succumb to malnutrition from climatic influences on food supplies (Barboza et al., 2009). Aggregate-female survival can be estimated from count data using

$$\frac{females_t}{females_{t-1} + (juveniles_{t-1} + sub\text{-}adult_{t-1})}. \qquad 4.1$$

Across 1998 to 2016, aggregate-female survival was high (mean = 0.865, $n = 19$) and did not display much variability (standard deviation [s] = 0.114).

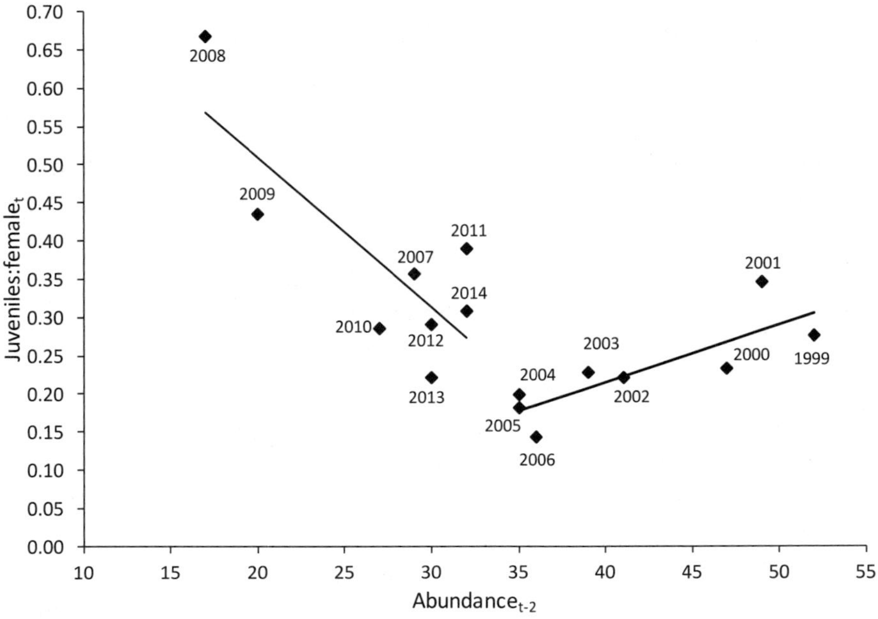

FIGURE 4.6. The estimated relationships between abundance of females, juveniles, and subadult males lagged 2 years and juvenile recruitment (juveniles:female) in the Davison herd between 1999 and 2014. The regression during the decline stage was between 1999 and 2006, and the regression during the lower *K* carrying capacity stage was between 2007 and 2014. The graph was originally published as Figure 3 in Starns et al. (2014).

The standard deviation expressed as a proportion of the mean (CV) was 0.13. Juvenile recruitment (juvenile:female) was much more variable because the CV was 0.39 (mean = 0.303, s = 0.117). If climatic conditions affected elk demography, it was more likely that density-independent factors influenced juvenile: female ratios. But, the findings of Starns et al. (2014) indicated no influence on juvenile recruitment from precipitation and temperatures in summer or winter. Recruitment appeared to be driven by density dependence.

One of the key findings from Starns et al. (2014) was that density-dependent juvenile recruitment differed between the decline and lower *K* carrying capacity stages (Fig. 4.6). Moreover, there was a 2-year lag in the response. The break between stages in that analysis used the year of the abundance nadir, 2006. During the decline stage there was a weak, positive relationship that indicated a correction in density-dependent recruitment, whereas during the lower *K* carrying capacity stage there was an inverse, linear relationship. That relationship is what I would expect if a new and lower *K* carrying capacity was developing. The inverse relationship demonstrated negative feedbacks between the food supply and herd abundance.

Viewing Figure 4.5, that density-dependent recruitment could have developed before the year of the nadir in abundance. In the latter part of the decline stage, food supplies should still be limited and females, which probably have depleted fat and protein stores, should favor their own survival and not risk investment in reproduction, which means juvenile:female ratios should be low (Sinclair & Parkes, 2008). Starns et al. (2014) considered a break in the relationship only in the year of the nadir. Demographic changes should not have to coincide with the exact year of the nadir when density-independent factors are not influential.

To further examine the year of the break in the relationship between herd abundance and juvenile:female ratios, and whether the lag with herd abundance was 1 or 2 years, another model selection analysis was conducted (Table 4.2). Nonetheless, the findings were similar to what Starns et al. (2014) reported. The addition of 2 more years of data did not change the patterns much. The model with a break in the relationship in 2006 had at least a slightly larger log likelihood and r^2 than a break in any of the other years considered. Perhaps the most notable change was that the relationship weakened between Starns et al. (2014) and this analysis (r^2 dropped from 0.71 to 0.61). The nadir year, 2006, was indeed the break year and juvenile recruitment was more strongly coupled to a 2-year, not a 1-year, lag with herd abundance. During the decline stage before the nadir, there was not a substantial relationship between abundance and juvenile recruitment. After the nadir, there was a classic, density-dependent relationship between herd abundance and juvenile: female ratios.

The inverse relationship between abundance and juvenile:female ratios after 2006 indicates that density dependence was influencing a demographic attribute of the Davison herd. In turn, density-dependence should affect the dynamics of abundance and the emergence of a new lower *K* carrying capacity (Fig. 4.7). The inverse relationship between abundance and juvenile:female ratios might shed some light on what might be the new lower *K* carrying capacity. If the regression is extended beyond the largest observed abundance to the point where it intercepts the X-axis, the estimate of zero recruitment can be used to project a new lower *K* carrying capacity. That projection is 54 (Fig. 4.7). Estimating *K* in this manner might result in an overestimate. What has not been factored in is female composition of the herd, sex ratio at birth, and the

TABLE 4.2. Summary of AIC_c model selection analysis to estimate relationships between abundance (females, juveniles, and subadult males) and juvenile:female ratios in the Davison herd. Least squares regressions were built to assess whether abundance lagged 1 (N1) or 2 years (N2) influenced juvenile:female ratios and whether relationships changed at or near the nadir (2006). Possible breaks in relationships were 2004 (break4, separate regressions for data from 1999–2004 and 2005–2016), break5 (2005), so on, to break8 (2008). The selected model has an asterisk. Also provided were the r^2 for each model. The X denotes an interaction. The N1 and N2 models estimated three parameters and the remaining models estimated five parameters. Below are the parameter estimates for the selected model.

MODEL PREDICTORS	LL	Δ	r^2
N2	16.03	6.54	0.30
N2, break4, N2 X break4	21.62	2.64	0.54
N2, break5, N2 X break5	22.03	1.83	0.56
*N2, nadir, N2 X nadir	22.94	0.00	0.61
N2, break7, N2 X break7	20.09	5.71	0.46
N2, break8, N2 X break8	16.17	13.44	0.17
N1	14.80	9.00	0.20
N1, break4, N1 X break4	15.63	14.62	0.08
N1, break5, N1 X break5	16.17	13.44	0.11
N1, nadir, N1 X nadir	19.18	7.51	0.40
N1, break7, N1 X break7	18.84	8.21	0.38
N1, break8, N1 X break8	15.26	15.37	0.08

COEFFICIENT	ESTIMATE	SE	t	P
Intercept	−0.086	0.182	−0.47	0.643
N2	0.007	0.004	1.75	0.102
Nadir	0.869	0.220	3.95	0.001
N2 X nadir	−0.022	0.006	−3.65	0.003

length of time that females need to live to recruit enough females to maintain that abundance (McCullough, 1979). Because these attributes are not static across years but vary, I keep things simple. Until more years of abundance data are available to directly estimate *K* (Chapter 6), the tentative estimate of *K* is 54.

Now that there is an estimate of *K*, it is possible to make a crude prediction of the year when the abundance of the Davison herd should come into equilibrium at a new *K* carrying capacity (Boyce, 1989; McCullough, 1979). To make the prediction, I used the standard logistic equation and estimates of the maximum intrinsic rate of increase (r_{max}) and *K* from the time series between 2006 and 2016 (Fig. 0.3). Time lags can have profound impacts on population dynamics when they are lengthy and *K* carrying capacity is high (Caughley, 1976; Forsyth & Caley, 2006). Given the projected new lower *K* carrying capacity was not high and the time lag was just 2 years, I assumed the time lag would not have much of an impact on population dynamics. For r_{max}, I used

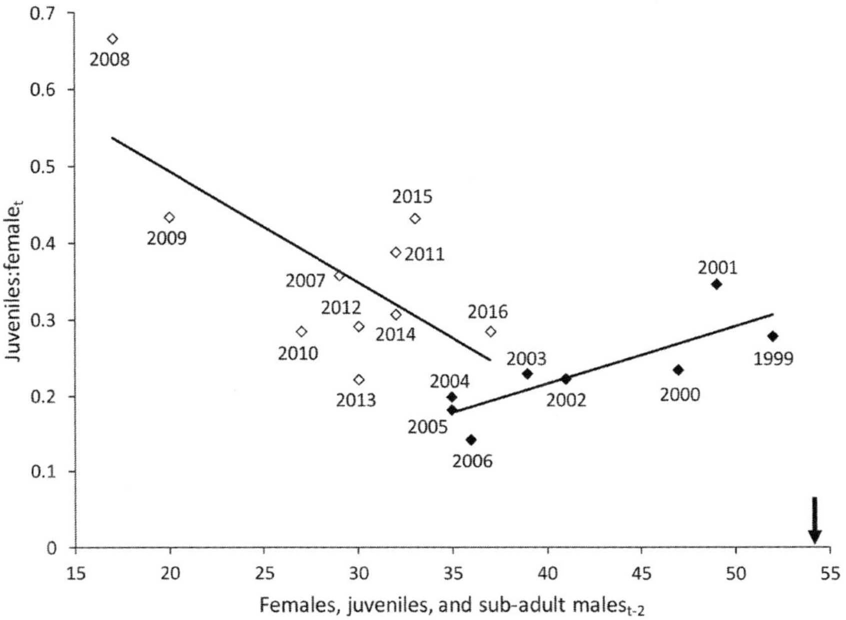

FIGURE 4.7. The estimated relationships between abundance of females, juveniles, and subadult males lagged 2 years and juvenile recruitment (juveniles:female) in the Davison herd between 1999–2006 and 2007–2016. The scatterplot is an update of what was reported in Figure 4.5. The vertical arrow along the Y-axis is the projected abundance where recruitment is zero (54) and, therefore, a projection of a new lower *K* carrying capacity.

population growth between 2006 and 2007 (0.30); these 2 years had the fastest growth since the nadir. The predicted time to reach 50 animals from 2006 was 11 years, which was 2016. Davison herd abundance was actually 51 in 2016 (Fig. 4.2). The population should reach the new lower *K* carrying capacity in about 14 years after 2006. That would make 2019 the year that the Davison herd reaches the new equilibrium. Collecting population data into the foreseeable future is needed to clarify the lower *K* carrying capacity of the Davison herd.

SURPLUS FOOD SUPPLY

What defines a population irruption is intense herbivory that alters plant communities and reduces ungulate food supplies. The most convincing evidence of an irruption, therefore, is a drop in food supplies that occurs across the overshoot, decline, and lower *K* carrying capacity stages. The bulk of the elk food supply throughout the year in the Prairie Creek drainage is meadow vegetation

(Harper, 1961). Because the Davison herd is nonmigratory and feeds throughout Davison Meadows, forage biomass needs to be measured throughout all parts of these meadows during all seasons and years of the irruption. Directly measuring food supplies in the field at that temporal and spatial scale can be done but it is time consuming and beyond what my shoestring budget allowed. An indirect assessment of vegetation characteristics can be obtained from infrared images taken by satellites. The indirect measure is the NDVI (Starns et al., 2015). The NDVI is a measure of plant greenness. Yet, this measure of greenness correlates with vegetation biomass and growth (Starns et al., 2015). Satellite images of NDVI at a resolution of 30 m^2 are available every 16 days, assuming there is no cloud cover, a perennial issue along the cloudy and foggy coast of northern California. Enough images, however, were obtained between January 1989 and December 2011 to estimate NDVI in the Davison Meadows, Boyes Meadows, and the Bald Hills Meadows (Starns et al., 2015). Images after 2011 were spotty because of satellite issues. If the Davison herd irruption affected the food supply, then NDVI attributes obtained from the Davison Meadows should differ before 1991 and afterwards. There should also be no changes in NDVI characteristics before and after 1991 in the two meadow complexes occupied by non-irrupting elk herds.

Attributes were the mean and standard deviation of NDVI across all the pixels in meadows in a month. A pixel encompassed a 30 m^2 area. The standard deviation of NDVI was expressed as a proportion of mean NDVI, the coefficient of variation or CV. The CV of NDVI is a measure of spatial variation. In Davison Meadows, changes in mean and CV of NDVI occurred before and after 1991 (Fig. 4.8). In September, near the end of the dry season, there was a drop in mean NDVI from 1989 to 1992 that flattened out in the years afterwards. In May, a month of peak forage biomass, there was a decline in CV of NDVI from 1989 to 2011 that was steeper earlier than later in the time series. These patterns were not detected in the Boyes and Bald Hills meadow complexes (Starns et al., 2015).

Patterns in NDVI were suggestive but not convincing evidence that a change in food supply was associated with the irruption in the Davison Meadows. The NDVI is a coarse measure of vegetation biomass and cannot detect changes in composition of plant communities in meadows. In grassland systems, vegetation differences associated with ungulate herbivory can be subtle. My instrument that measured vegetation attributes lacked the resolution to detect possible changes in vegetation associated with particular irruptive stages.

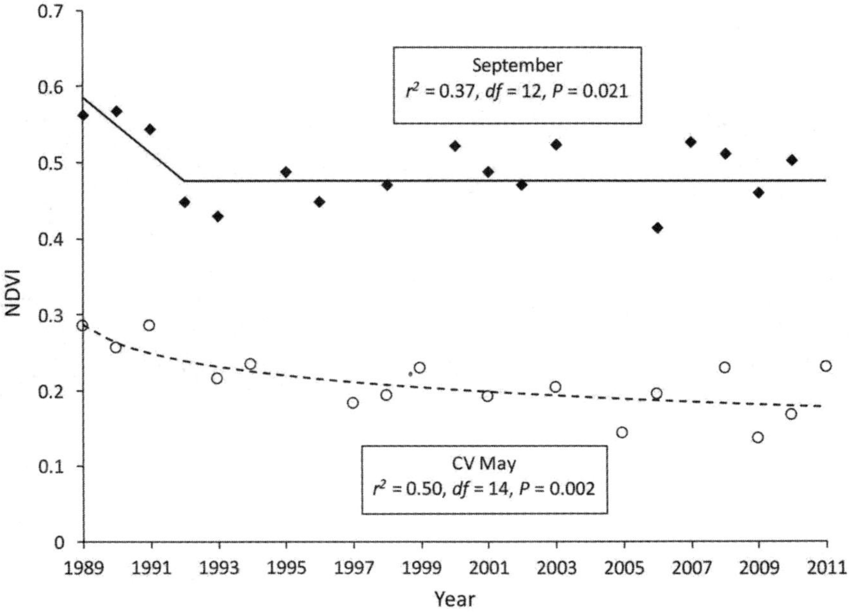

FIGURE 4.8. Relationships between year and September NDVI and CV of May NDVI for Davison Meadows from 1989 to 2011. The September NDVI relationship was estimated with a piecewise regression. The nonlinear relationship estimated for CV of May NDVI was from natural logarithmic transformation of both variables. The relationships were originally published in Figures 5 and 6 in Starns et al. (2015).

What the temporal patterns in NDVI do indicate was that elk herbivory in Davison Meadows in September was associated with a drop in vegetation biomass between 1989 and 1992. The spatial variation in NDVI (CV) also seemed to decline from 1989 to 2011 at the peak of the growing season in the Davison Meadows. Vegetation characteristics became less, and not more, heterogeneous across these years.

One change I observed in the Davison Meadows across the years of my study was the ostensibly slow but steady displacement of grasses in parts of Davison Meadows by one species, reed canary grass (*Phalaris arundinacea*) (Fig. 4.9). Reed canary grass is an invader of open, moist areas that are common in the Davison Meadows (Starns et al., 2015). Once reed canary grass becomes established, it can displace other plant species by the plasticity of its clonal architecture (Green & Galatowitsch, 2002; Martina & Von Ende, 2013). Elk, furthermore, do not appear to readily consume reed canary grass. This might be because of the alkaloid concentration in reed canary grass (Hagman,

FIGURE 4.9. Photographs of similar views of sector B in Davison Meadows in August 1996 and July 2013. Note the increase in prevalence of reed canary grass. The vertical arrows are pointing to the same telephone pole in both photographs. The photographs were reprinted from appendix S1 in Starns et al. (2015).

Marten, & Hovin, 1975; Marten, Barnes, Simons, & Wooding, 1973; Woods & Clark, 1971). Alkaloid concentration in plants is inversely related to herbivore consumption (Starns et al., 2015). The observed increase in reed canary grass in Davison Meadows from 1996 to 2013 might be facilitated by light grazing pressure by elk and heavier grazing pressure on other, palatable species of grasses and forbs. The surplus food supply during the rapid growth stage seems to be palatable species of grasses and forbs that were displaced by less palatable species such as reed canary grass, a grass resistant to elk herbivory.

INTERPRETATION

Ungulate irruptions can alter plant communities, reduce food supplies, and affect ecosystem processes (Holling, 1973; Leopold, 1949; Ricca, 2013). Reductions in food supplies might erode elk resilience owing to a lower *K* carrying capacity. The conditions for an irruption appear to be ideal on many oceanic islands. There is a rich food source that is unable to adjust to ungulate herbivory, with no predators to check rapid population growth and nowhere for animals to go once the food supply is exhausted. These conditions were not present in

Davison Meadows in 1991 when elk were given unrestricted access. The unique setting was useful for examining the drivers of the elk irruption.

My results demonstrate that the first stage, rapid population growth, of the irruption was facilitated by elk moving into and then remaining in a new, rich food source, the Davison Meadows. Adult males were most numerous during the overshoot and into the decline stage. Males might have been drawn to Davison Meadows during the rapid growth stage because of the food supply and potential mating opportunities with females.

Discussion on precursors to rapid growth and overshooting the food supply has focused on food supplies not exposed to ungulate herbivory, predators, climate, and limits to movement (Ricca, 2013; Tyler, 2010). The enormously high r_{max} displayed by the total population during the rapid growth stage, presuming colonization by two elk, suggests that adult male immigration helped fuel rapid growth. In addition to the identified precursors, the age–sex structure of animals available to colonize an area might also have played a role.

The decline after the overshoot stage was gradual and took place over 10 years. There was no sudden die-off of animals and female survival was high. There also were no density-independent influences detected on juvenile recruitment. Density-independent factors such as climatic extremes probably had little influence on the dynamics of the Davison population irruption. A die-off precipitated by density-independent deep snow and thick ice, for example, means a drop in vital rates coincides with a drop in abundance. The stages of irruptions delineated by abundance are coupled to demographic processes in that stage. When density dependence is the primary influence on juvenile recruitment, then demographic processes should be able to extend across irruptive stages. Yet, density-dependent recruitment in the lower *K* carrying capacity stage appeared to coincide with transition from the decline to lower *K* carrying capacity stage identified from the time series of herd abundance.

While I have talked about the lower *K* carrying capacity stage, my data do not allow me to predict an estimate of *K* with confidence. Using the total population size to estimate *K* is unreliable because processes affecting numbers of males and females (including juveniles and subadult males) do not appear to be the same across the irruptive stages. For males, I am unable to identify if the drop in their numbers across the time series is due to food limitations in Davison Meadows or if males simply moved to Boyes Meadows because it was an unexploited forage habitat. Males are typically less gregarious than females and might occupy Boyes Meadows to reduce encounters with other elk in the more populated Davison Meadows (Weckerly, 2001). Male forage intake, compared to females, seems to be more adversely affected by elk in close proximity (Peterson, 2016). Another complication is that Davison herd abundance has yet to arrive at an equilibrium with its food supply. Between 1997 and 2006 there were no density-dependent relationships detected. After 2006, juvenile recruitment appeared to be density dependent, indicating that food limitations were affecting the herd. Nevertheless, up to 2016, herd abundance did not appear to be affected by food limitations. In a long-lived species like elk such a pattern can happen. Adult females are buffered from environmental vagaries, have a high survival rate, and persist for years. Recruitment of juveniles, however, is more variable and likely to be sensitive to environmental vagaries such as food limitations. It might sound like a ploy to keep studying or chasing elk, yet more data are needed to estimate *K* carrying capacity of the Davison herd.

The extent of the drop between total abundance at the peak of the overshoot (83) and the total abundance in 2016 (51) was not nearly as dramatic as what probably occurred with reindeer on St. Matthews Island. Implying that the extent of change in food supplies and plant communities in Davison Meadows was also not as dramatic as what occurred on St. Matthews Island (Klein, 1987). The change in the Davison food supply was ostensibly subtle from year to year. Palatable species of grasses and forbs, the likely surplus food supply, were gradually displaced by less palatable species such as reed canary grass.

Given that elk irruptions do occur in the RNSP, what can be done to prevent irruptions in the future? Identifying when an irruption is most likely to happen needs to be clarified, however, to have a high probability of conducting an action that prevents irruptions. I harp on identifying the environmental conditions conducive to an irruption because most ungulate irruptions, including this one, were not anticipated. A surplus food supply that facilitates rapid population growth might not be readily apparent. In my view, irruption of an existing herd in the RNSP is unlikely simply because most other extant herds are already near *K* carrying capacity (this chapter and Chapter 6). The most likely scenario for another irruption is through the colonization of new forage habitat by elk. Nonprofit institutions such as Save the Redwoods League are purchasing private lands adjacent to the existing RNSP boundaries and deeding the land to the RNSP. These lands often do not have existing elk herds and the habitat can be closed-canopied forests, retired lumber mills, or cattle pastures. If cattle pastures or lands restored to meadow are 20 ha in size or larger, comparable to the size of the Davison Meadows, then the opportunity for an irruption seems possible. A new forage habitat suddenly becomes available and elk in the vicinity immigrate and colonize the area with a rich, new food source. The new herd grows rapidly, setting up the conditions for the remaining irruptive stages. The purchase of new lands with 20 ha or more of meadow habitat that do not already have a herd are possible candidates for an elk irruption.

It always seems that female and male differences in spatial patterns and resource use require some understanding to explain drivers of population dynamics in ungulates (Clutton-Brock et al., 1982; McCullough, 1979). My study is no different. Females and males consume similar forage plants in meadows in the RNSP (Harper, 1961). Yet, habitat use and spatial patterns differ between females and males. Both females and males used meadows more than 50 percent of the time but males used forest more often (Bliss & Weckerly, 2016). Early in the study, when abundance was high, females and males partitioned Davison Meadows. Some parts were used almost exclusively by females and other parts by males (Weckerly, 1998; Weckerly et al., 2004). Later in the study, when abundance was lower, males ceased being prevalent in Davison Meadows. Early on, there was also sexual segregation that changed to females and males not aggregating or segregating in meadows, a random pattern (Peterson, 2016). These patterns were in winter. Sexual segregation and intersexual resource partitioning is usually greatest during parturition in late spring and early summer (Stewart, Walsh, Kie, Dick, & Bowyer, 2015). It is plausible to expect that sexual segregation in the Davison herd also occurred during parturition. These differences between the sexes in habitat use and spatial patterns might then explain resource partitioning. The consequence of intersexual resource partitioning was that herd abundance, not total abundance, influenced juvenile recruitment, a pattern evident in other ungulate populations. What is tantalizing to me is to consider whether the abundance of males might have negatively influenced herd abundance at the peak of the overshoot (Weckerly, 1998). At that time, the Davison herd was not using all parts of Davison Meadows. The parts of Davison Meadows that were not used by the herd were occupied by males. If male abundance had been less during the overshoot period of the irruption, males should then have had lower grazing pressure in Davison Meadows. The result might have been more forage for the Davison herd, higher juvenile recruitment, and a more abundant herd than what I observed during the overshoot period.

In summary, the Davison irruption had the elements that define an ungulate irruption as well as novel features. There were the characteristic four stages of the irruption and a surplus food supply. To understand what prompted the irruption, however, required awareness of what was occurring at multiple spatial scales. Stated another way, the irruptive dynamics in the Davison Meadows probably depended on the elk distribution, age–sex composition, and population dynamics in areas beyond the Davison Meadows. How the Davison irruption manifested and appears to be playing out probably differs from irruptions on oceanic islands. Be that as it may, my findings are consistent with the quote by Caughley (1976) at the beginning of this chapter. When ungulates have a surplus food supply, an irruption is likely.

5

The Unexpected

Population Extinction

Hustlers grab your guns.
Your shadow weighs a ton, driving down the 101.

— Phantom Planet, "California"

I was surprised to witness the local extinction (simply referred to as extinction hereafter). The documentation on the Boyes herd indicated that it had been around longer than most herds in the RNSP and its abundance seemed to be high enough such that there was no discussion about extinction (Harn, 1958; Kristan, 1992; Lieb, 1973). There was no hunting or other obvious anthropogenic activity that could be fingered as precipitating the demise of this herd. Furthermore, Boyes Meadows was burned periodically between 1997 and 2010 and additional forage habitat was created with the completion of the Bypass in the early 1900s (Chapter 1). The burning should have maintained or even increased elk forage and its nutritional value; the completion of the Bypass might have increased *K* carrying capacity (Kolbe, 2016).

A few elk were first documented in Boyes Meadows in 1937; by the late 1940s the Boyes herd probably numbered over 100 elk (Fig. 1.3). From the early 1950s to the late 1990s, the herd's abundance fluctuated between 20 and 60. When I began population surveys in 1997 the herd size was 29 (Fig. 5.1). In 1998 the herd increased in size by one, to 30. After 1998 there was a gradual decline until 2006, when five elk composed the herd. From 2007 to 2009 herd size was five and in 2010 it was four. Thereafter no females, juveniles, or subadults were detected during January surveys. Another unusual feature was that across 1997 to 2010, juvenile recruitment failed to display a classic, inverse density-dependent response (Fig. 5.1). As abundance went down, there were years when the rate of juvenile recruitment was high, as expected, but there

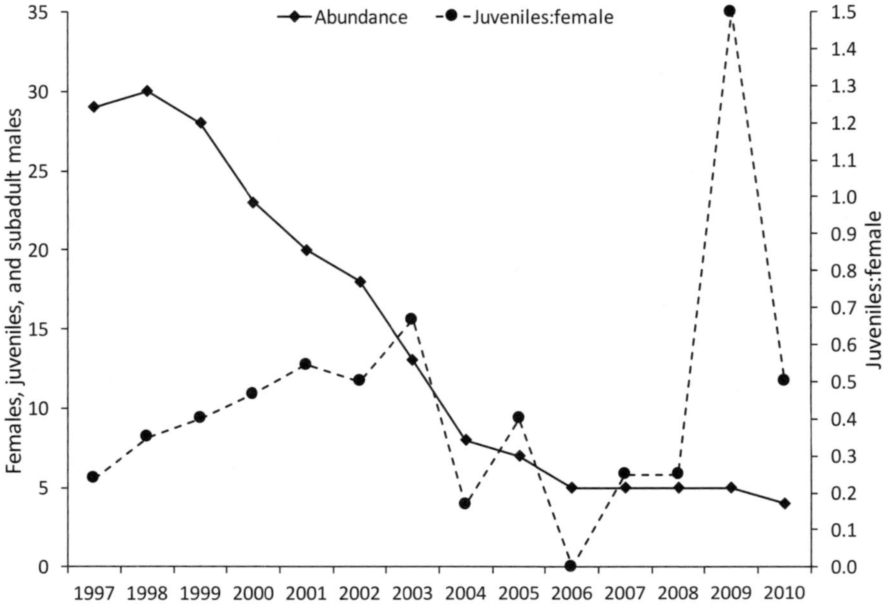

FIGURE 5.1. Abundance and juvenile recruitment estimates for the Boyes herd, 1997–2010. After 2010 no females, juveniles, or subadult males were detected during population surveys in January. The relationships between abundance and juveniles:female were not statistically significant at lags of 1 ($F = 0.009$, $df = 1, 10$, $P = 0.925$) or 2 years ($F = 0.045$, $df = 1, 9$, $P = 0.836$). There was no density-dependent juvenile recruitment detected across these years.

were also years when it was low. The lack of an inverse density-dependent response in juvenile recruitment indicates that food limitations were not a plausible culprit to explain the demise of the Boyes herd.

Extinction of the Boyes herd needs clarification. Extinction is the opposite of persistence. Before 2011 there was a herd that used Boyes Meadows and surrounding areas year-round. Extinction is what has occurred since 2011: no females, juveniles, or subadult males have been detected in Boyes Meadows or the Bypass during January population surveys. As will be presented shortly, these two areas were the primary areas used by the Boyes herd. Adult males do use Boyes Meadows and the Bypass, and they have been present in numbers between three and 10 from 2011 to 2016 (Starns et al., 2015). Across those same years, one to two females with one or two juveniles were also observed in Boyes Meadows from the season of parturition until September and sometimes into November. Boyes Meadows and the Bypass is not forage habitat vacant of elk—it is forage habitat without a herd using these meadows in all seasons of the year.

In populations of large herbivores that persist, the demographic variables that drive change in abundance were mentioned in the previous chapter (see Raithel, Kauffman, & Pletscher, 2007). Adult female survival is high and does not display much variation from year to year. Production of juveniles and their survival in the first year of life, in contrast, varies across years. Variability in production and first-year survival of juveniles is because females appear to favor their own survival over the risks involved in producing and provisioning young (Gaillard & Yoccoz, 2003). These life history patterns result in variation in population growth being coupled to variation in juvenile recruitment from year to year. Yet, the greatest effect on population resilience stems from variation in adult female survival. Populations that persist most likely do so because females have consistently high survival rates across years.

The aforementioned patterns in demography indicate that extinction is probably because of recruitment failure, low female survival rates, or both. Recruitment failure is when there is no recruitment or the level of recruitment cannot replace adults that are lost through mortality. Due to the relatively high survival rate of adult large herbivores, an increase in mortality might result in female survival rates declining from, say, 85 percent annually to 70 or 60 percent (Raithel et al., 2007). The outcome of a drop, but not a precipitous drop, in annual survival rate is that the decline takes place over a number of years. The herd decline will reach a size where the benefits of grouping lessen and mortality increases for remaining herd members (Fryxell et al., 2008). A small group size does not seem to offer much in terms of detecting and diluting predators. Knowledge about resource distribution and abundance might also erode. When herds reach a small size, Allee effects manifest and the herd dies out or the remaining herd members move on to other areas (Hutchings, 2015). Either way, extinction in the area that used to be occupied by the herd is the result.

The decline in the Boyes herd from 1997 to 2010 does not appear to be from a lack of young being recruited into the herd (Fig. 5.1). There was only 1 year (2006) when no young were recruited. From 1997 to 2003 recruitment gradually increased, which coincided with a decline in herd size. After 2003 recruitment appeared more erratic. Erratic recruitment is related to fewer numbers of females in the herd and the large difference in the juvenile:female ratios that occur when there is a difference of just one juvenile. In 2009 the juvenile:female ratio seemed to be impossibly high at 1.5. Perhaps a mother died after the young had been weaned but before my surveys in January. It is conceivable that a female gave birth to two young. Yet, a litter of two seems to occur so rarely (less than 1 percent) that I ignored the possibility (Church & Hudson, 1996).

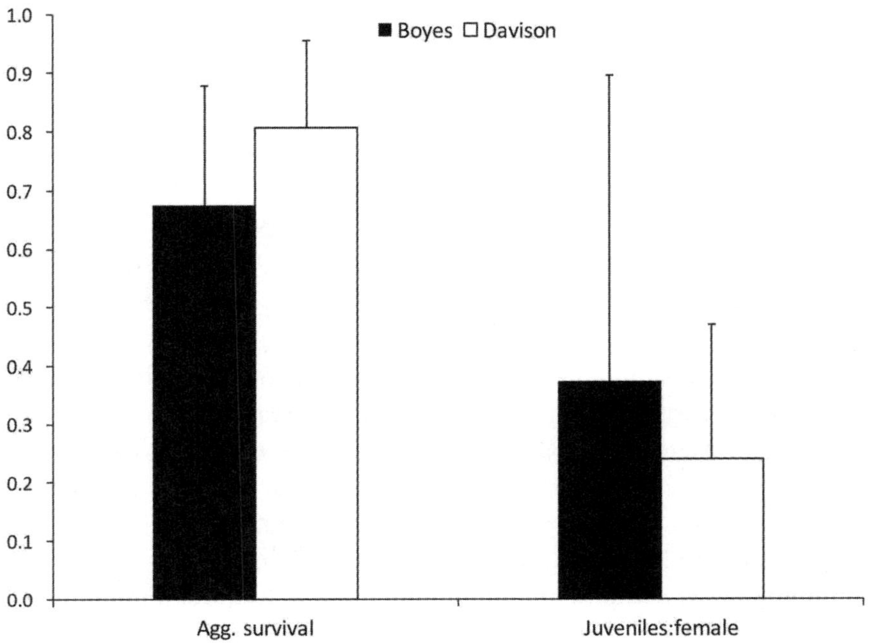

FIGURE 5.2. Means and coefficients of variation (error bars) of aggregate survival rate of females and juvenile recruitment in the Boyes and Davison herds, 1997–2006. Across these 10 years both herds declined. Differences between herds in means and process variances were assessed with randomization tests. Tests were conducted on 10,000 random permutations of the data. Davison aggregate survival rate was greater than Boyes aggregate survival rate ($P = 0.0172$) but the variance in survival rate was greater for the Boyes herd ($P = 0.0129$). Juvenile:female ratios were greater, on average, for the Boyes herd ($P = 0.0245$), as was the variance in the ratios ($P < 0.001$).

Coincidentally, the nearby Davison herd declined across 10 of the same years as the Boyes herd, 1997–2006 (Fig. 4.2). Unlike the Boyes herd, the Davison herd did rebound and exists as of 2016. Comparisons of female survival and juvenile recruitment between the herds across those years might be useful to testing whether those demographic variables were low in the Boyes herd. Mean female survival rate (Formula 4.1) did appear to be higher in the Davison herd than in the Boyes herd (Fig. 5.2). Perhaps female survival was low in the Boyes herd from 1997 to 2006. The mean juvenile:female ratio, however, was higher in the Boyes herd than in the Davison herd. This finding is not too surprising, given the inherent variability of juvenile recruitment in large herbivores.

A herd that goes extinct might also display greater variation in demographic variables (Akçakaya, 2002; Kendall, 1998). Variation in female survival

TABLE 5.1. Variance estimates for aggregate, female survival rates, and juvenile:female ratios of the Boyes and Davison herds, 1997–2006. In these years both herds declined. Variances were estimated using formulas presented in Akçakaya (2002) and Kendall (1998) for count data. Use of these formulas assumes observation error is trivial or nonexistent.

	AGGREGATE SURVIVAL		JUVENILE:FEMALE	
VARIANCE	BOYES	DAVISON	BOYES	DAVISON
Process	0.019	0.011	0.020	0.009
Demographic	0.011	0.004	0.017	0.007
Environmental	0.008	0.007	0.003	0.002

and juvenile:female ratio was greater in the Boyes herd (Fig. 5.2). Greater variability in herds that go extinct might be because of greater environmental stochasticity, demographic stochasticity, or both (Kendall, 1998). Stochasticity refers to a variety of factors, some of which might be known, some of which might be unknown. Stochasticity is often measured as a variance. Environmental variance could be from climatic and elk effects on food supplies, predation, disease, poaching, collisions with vehicles, and any other extrinsic factors that might have affected all elk in a herd (Caughley, 1994). Demographic variance captures the stochastic events that can affect survival and productivity of individuals in a herd. The two variances are independent of one another. Furthermore, when abundance is small, demographic variance is usually greater than environmental variance. The total variance, labeled process variance, is the sum of demographic and environmental variances.

In the Boyes herd, the respective process variances in female survival rate and juvenile:female ratio were 1.7 and 2.2 times greater than in the Davison herd (Table 5.1). The larger process variances in the Boyes herd were mostly because of demographic variance. Demographic variance in the Boyes herd was at least two times larger than demographic variance in the Davison herd for both female survival rate and juvenile:female ratios. Environmental variance appeared to be similar between the herds for both demographic variables. The greater demographic variance in the Boyes herd might be because of small abundance (Akçakaya, 2002). The effect of the demise of one female on female survival rate or the loss or addition of one juvenile on juvenile:female ratios is more dramatic at low abundance.

Identification of agents that might cause herd extinctions would be useful information to RNSP biologists. The possible agents that enable a population decline to continue to extinction are very numerous and the list is growing in the contemporary world (Caughley, 1994). Agents are factors that reduce food supplies or prompt mortality that might be difficult for animals to perceive or mitigate. Agents can be the usual suspects: habitat degradation, loss and fragmentation, pathogens and diseases, invasive species that are effective competitors or predators of native species, poaching, and so on. In the contemporary world, agents are usually connected to anthropogenic activities.

The spatial pattern of female and juvenile mortalities could be useful to fingering potential agents that contribute to the demise of the Boyes herd. Mortalities were apparent mortalities. A mortality was definitive when I located a carcass and was able to identify whether the animal was female or juvenile. I ignored subadult males because of the possibility of dispersal. If an RNSP biologist, park ranger, or state police officer reported a mortality that I did not view, then I informally interviewed the person to evaluate the veracity of his or her statements. The apparent in apparent mortality was when the count dropped by one female or juvenile between 2 consecutive days. Subsequent counts were also minus an elk. Between the two consecutive counts, I did not locate a carcass but because of the group locations I was able to assign the location of the apparent mortality to either Boyes Meadows or along the Bypass (Fig. 5.3). Counts were during population surveys in January and at other times in the winter and early spring (see below). Between 1997 and 2010 I detected 20 apparent mortalities of females and juveniles from the Boyes herd (hereafter mortalities). A crude measure of the total number of mortalities across these years can be obtained by taking the sum of females and juveniles in year t minus the sum of females and subadult males in year $t + 1$. The measure, however, cannot detect juvenile mortality between time of birth and January when I conducted the counts. Nonetheless, this measure indicated that there was a minimum of 60 mortalities in the herd. The 20 mortalities I accounted for was 33 percent of those 60 mortalities. In reality, that percentage was probably lower. Assuming my data on locations of mortalities are unbiased, 75 percent of mortalities occurred along the Bypass and the remaining 25 percent occurred in Boyes Meadows.

Estimating elk use is required to assess the risk of mortality along the Bypass. Unfortunately, estimating Boyes Meadows and Bypass use by the Boyes herd across all months or seasons from 1997 to 2010 was beyond my shoestring budget. A particular drawback was that I was unable to estimate use in all seasons in the years when much of Boyes Meadows was burned. In September 1996 and 2000, all of Boyes Meadows was burned and in September 2005, 2006, and 2008, parts of Boyes Meadows were burned. The burning appears to increase meadow use in January and probably in other months as well (Chapter 2). Despite that, in 5 years between 1995–1996 and 2001–2002 I located the

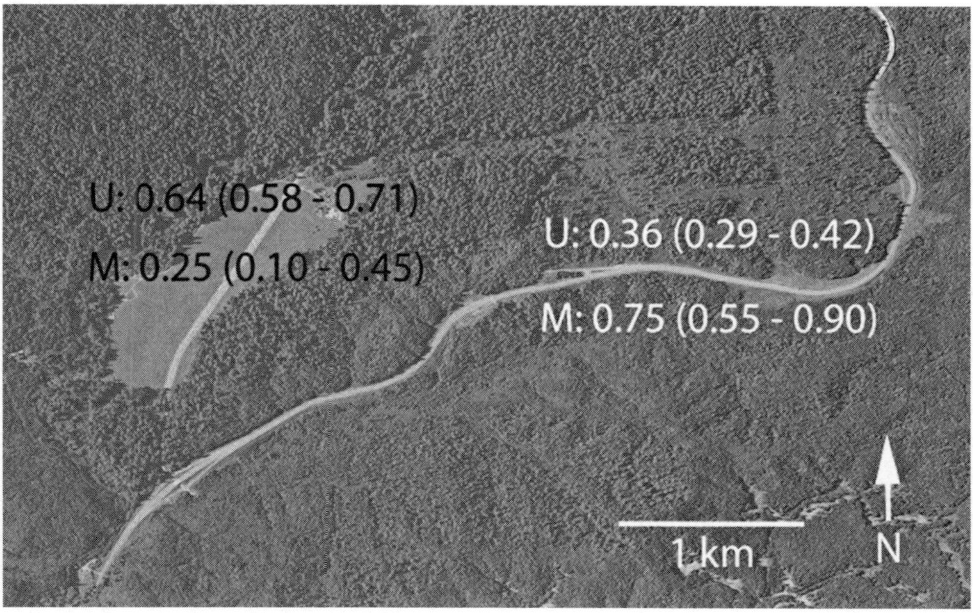

FIGURE 5.3. Aerial photograph of Boyes Meadows and the Bypass, U.S. Highway 101, to the south and east of Boyes Meadows. The Bypass was built in the late 1980s and officially opened in 1992. The white print is the estimated proportion of female, juvenile, and subadult male carcasses or mortalities (M) located along the Bypass and the estimated proportion of time that the Boyes herd used (U) the Bypass from December to April in 1995–1996 (n = 31), 1996–1997 (43), 1999–2000 (65), 2000–2001 (32), and 2001–2002 (22). In parentheses are 95 percent CI obtained from 10,000 bootstrapped samples. Sample sizes are 193 locations and 20 mortalities. The same statistics are provided for Boyes Meadows in black print.

Boyes herd 193 times in the morning between December and April (Fig. 5.3). Across that time the Boyes herd used the Bypass 36 percent of the time; the remainder of the time the herd was located in Boyes Meadows. Across the 80 population surveys conducted between 2003 and 2010, the Boyes herd was located on the Bypass 30 percent of the time (95 percent CI: 0.10–0.40). The Boyes herd did use the Bypass a noticeable part of the time.

If mortality was more likely along the Bypass than in Boyes Meadows, what were the potential benefits to elk from using the Bypass? One obvious possibility is forage. Boyes Meadows is about 51 ha in size and meadows along the stretch of the Bypass used by the Boyes herd total 19 ha. Total amount of foraging area is greater in Boyes Meadows. Yet, per hectare of meadow, forage in January could be noticeably greater along the Bypass than in Boyes Meadows in some years (Fig. 5.4). Assuming that forage biomass has a substantial,

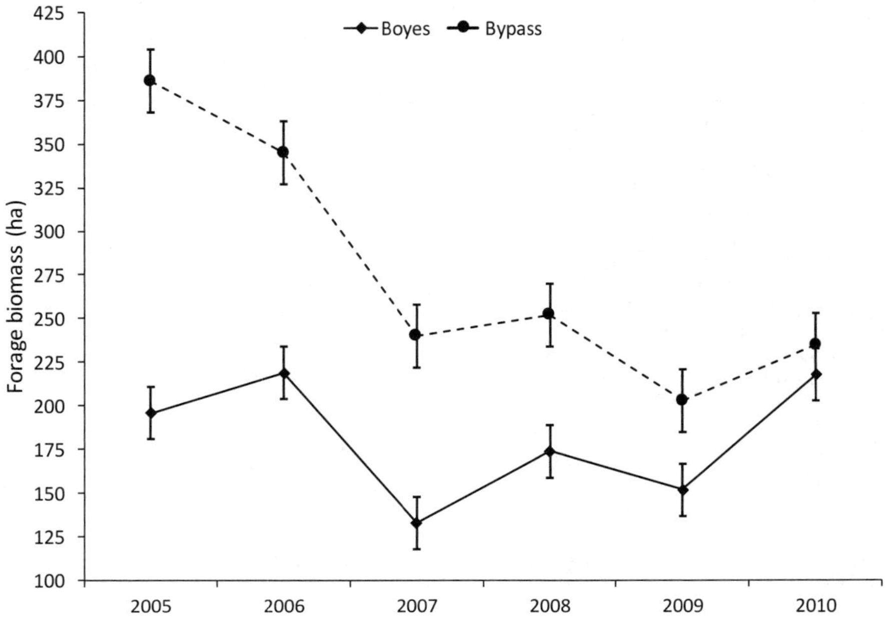

FIGURE 5.4. Forage biomass (per hectare) in Boyes Meadows and along the Bypass in January, 2005–2010. Forage biomass was estimated each year in 240 quarter-m^2 plots in Boyes Meadows and 60 quarter-m^2 plots in meadows along the Bypass. Means and standard errors were estimated with a linear mixed effects model with fixed factors of year, meadow (Boyes, Bypass), and a year X meadow interaction. There was a significant year X meadow interaction ($F = 12.69$, $df = 5, 1783$, $P < 0.001$). Transect was the random factor.

positive effect on bite size and forage intake in January, elk should be motivated to use the Bypass at that time of the year.

What evidence I have been able to cobble together suggests that the Bypass is bad for elk because of greater mortality. This inference can be stretched to infer that U. S. Highway 101 is bad for elk. At some level this inference is correct because elk can and do perish when they collide with vehicles and are poached, which are the two most likely culprits of mortality. The question then is, Does the level of mortality from these two sources exceed what herds can replace through recruitment and, perhaps, immigration? For the Boyes herd, the answer appears to be no. But there are other elk herds that inhabit areas that include the 101 highway.

Ten kilometers south of Orick the 101 goes through meadows near Stone Lagoon, which also has the Stone Lagoon herd. Most RNSP biologists are stationed out of the SOC in Orick but live in towns south of Stone Lagoon.

Sometimes RNSP biologists drive alone but most days they carpool to and from work. Twice a day during the work week, they pass through Stone Lagoon on the 101. There is no other route that RNSP biologists can drive. Between January 26, 2011, and April 19, 2012, these commuters recorded whether or not they sighted the Stone Lagoon herd while driving by. When sighted, the herd could be on the road, next to the road, or more than 100 m from the road. There was a total of 914 drives through Stone Lagoon and the herd was sighted on 60 percent of the drives. That sighting percentage is probably an underestimate. On 40 occasions, two vehicles passed through Stone Lagoon within five minutes of one another. Assuming the Stone Lagoon herd was either out of view or within sighting distance of the 101 during that five-minute interval, there were 12 times when the occupants in both vehicles did not agree on the sighting. At least one person in one vehicle saw the herd but no one in the other vehicle saw elk. It was likely that sighting elk while driving by was distorted by imperfect detection. Herd use of areas near the 101 during daylight hours was probably greater than 60 percent.

The opportunity for elk from the Stone Lagoon to collide with vehicles and to be poached from the 101 was there. Across the time that data were recorded, at least one elk did collide with a vehicle and perish. Yet, the Stone Lagoon herd appeared to coexist with the 101 (Fig. 5.5). In 2000, herd size was 87 and declined to 49 in 2003. Since 2003, herd size has gradually increased to the mid-60s. The prevailing evidence indicates that the Stone Lagoon herd is persisting in spite of the 101.

Another consequence of the shoestring budget was my inability to ascertain the source of mortality for all animals that perished along the Bypass. The two likely culprits, vehicle collisions and poaching, were the same as along the 101 through Stone Lagoon. Meadows were often a narrow strip next to the shoulder of the Bypass. Elk did graze these shoulders, which certainly increased the prospects for collisions with vehicles and made elk easy prey for poachers (Fig. 5.6). Also, stretches of the Bypass had grades and median barriers. Down-slope vehicles, which were southbound, might have had difficulty slowing down when they encountered elk. Median barriers could also restrict the mobility of elk attempting to evade and avoid vehicles. The median barriers were installed in 2002 and 2003. Coincidentally, the greatest decline by the Boyes herd ($\ln[N_t/N_{t-1}]$) was in 2003 (−0.49). Across the 13 years that I counted the Boyes herd, the probability of that occurring by random chance was 0.077 (1/13). Poaching is more challenging to document. I suspect poaching because it has been documented by park rangers along the Bypass and the

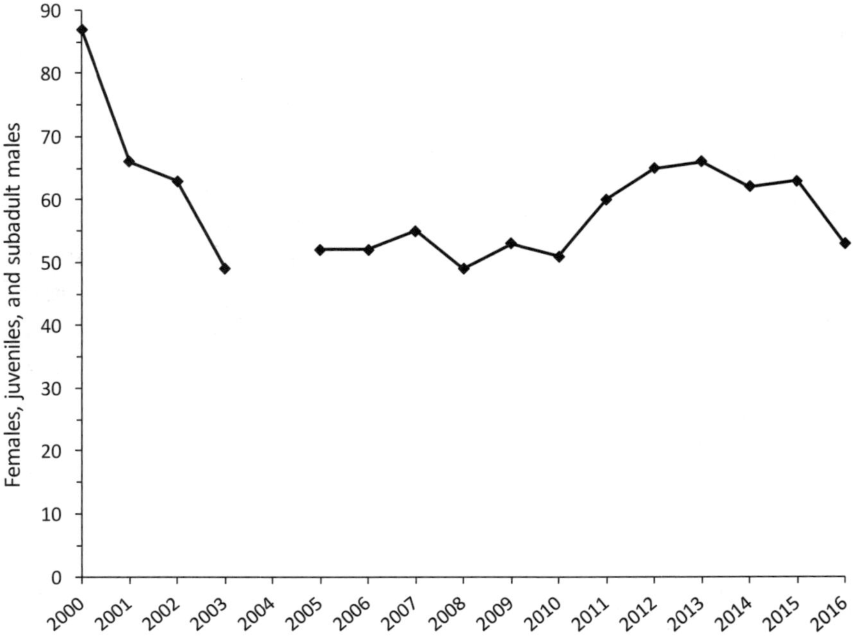

FIGURE 5.5. Abundance estimates of the Stone Lagoon herd, 2000–2016. The Stone Lagoon herd inhabited a meadow complex bisected by U.S. Highway 101 approximately 10 km south of Orick, California.

Bypass is a relatively remote stretch of the 101. There are no residences, convenience stores, gas stations, or campgrounds along the Bypass. Poachers might drive the Bypass during daylight hours, locate elk, and come back at night and shoot elk. The chances of getting caught poaching on the Bypass were probably less than along other stretches of the 101 that had human habitat. Unfortunately, I do not have robust evidence about the agents of extinction of the Boyes herd. But I will say that the demise of the Boyes herd was associated with the Bypass.

That a herd was observed to the point of extinction should provide information helpful to detecting potential extinctions in other herds in the RNSP. Abundance at *K* carrying capacity or the highest observed abundance is often the attribute examined in the study of population extinctions (Caughley, 1994; Hutchings, 2015). The less abundant the population, the greater the chance that it goes extinct from demographic variation. When I began population surveys in 1997, the Boyes herd was not abundant and thus the herd's initial abundance was cause for concern. But there were other herds in the RNSP that

FIGURE 5.6. Photograph of median barrier that was installed on part of the Bypass in 2003. Note the elk from the Boyes herd foraging on both sides of the highway near the shoulders. The picture was taken on January 14, 2005.

had abundances comparable to the Boyes herd in the late 1990s (high 20s, low 30s) that persisted (Fig. 3.10, and Chapter 6). Small population size is certainly cause for concern. But herd abundance in itself does not provide much practical insight into whether a herd will go extinct in the next 5 to 10 years.

A strategy to reduce population extinction is to detect a population decline that continues to extinction earlier rather than later. That way actions can be taken to alleviate herd demise before extinction is eminent. One method commonly discussed in conservation science is to detect declines in abundance using commonly applied statistical techniques such as regression (Westcott, Fletcher, McKeown, & Murphy, 2012). Year is the predictor and abundance the response variable. When a test of a slope, for example, indicates a negative slope, a population decline is real. The statistical power or the number of years needed to detect decline of a regression slope will depend on the steepness of decline, variability from year to year in abundance, and the precision of the population survey method. In the real world of field biology, the number of years that it takes to statistically detect a decline is often more rather than few. A trend analysis, consequently, takes years of data to detect a decline. Yet, the trend analysis informs little if anything about whether the decline portends extinction or merely a decline that populations experience.

Another possibility is viable population monitoring (Seavy et al., 2009; Staples, Taper, & Shepard, 2005). Viable population monitoring uses a population growth model, estimate of process variance (sum of environmental and demographic variance), and population size at time *t* to estimate probabilities of change in $t + 1$ or longer time intervals. Viable population monitoring can detect population declines that are more likely to continue to extinction quicker than the regression analysis described in the previous paragraph. Unfortunately for me, viable population monitoring is probably most informative when there is a longtime series of abundances, not the 13-year time series I had for the Boyes herd.

Determining the agents of decline should also provide information that might indicate whether a decline leads to an extinction. From a practical view, determining what kills animals and where they are killed requires frequent surveys. For a species in the RNSP such as elk that are not endangered, threatened, or even a species of conservation concern, the resources to conduct intense surveying are not available. This option is unrealistic. A more feasible option is to find indicators in the observed population data that are collected each year, indicators that can be readily calculated and emerge early rather than later in the decline.

Data from the Boyes herd indicate that pervasively low recruitment, recruitment lower than in the nearby Davison herd, is unlikely. Indicators of population extinction might be more evident in adult survival rates (Fig. 5.7). To explore demographic signals to extinction and how quickly they might emerge, I calculated means and coefficients of variation (CV) of female survival rate during decline years with 4 years of data (1997–2000), 5 years of data (1997–2001), so on, to 10 years of data. The CV was the square root of process variance divided by survival estimate. I calculated statistics for both the Boyes and Davison herds. Patterns were evident from as little as 4 years of data. Aggregate survival rate was greater than 0.8 across all years of pooling data for the Davison herd, but it was less than 0.8 for the Boyes herd. Coefficients of variation in survival rates for the Boyes herd were often twice those of the Davison herd. When mean female survival declined there was a corresponding increase in variability.

Based on the analyses, an indicator of population decline that might continue to extinction is a female survival rate at or below 0.80 and a CV of 0.18 or more. Yet, treating a female aggregate survival rate of 0.80 or less and a CV of at least 0.18 as thresholds that means extinction is imminent is unwise. They

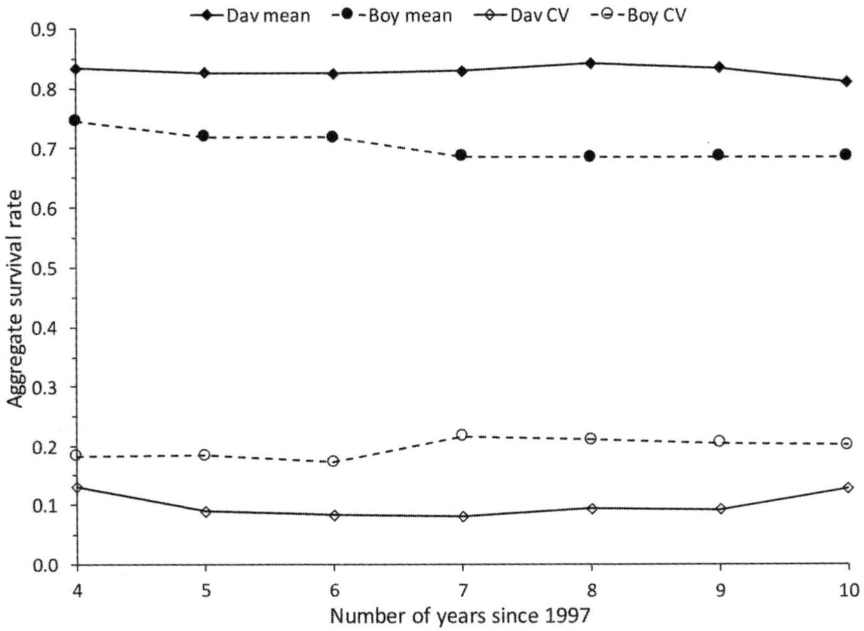

FIGURE 5.7. Line chart illustrating differences in means and variances (coefficient of variation – CV) between Boyes (Boy) and Davison (Dav) herds when 4 years (1997–2000), 5 years (1997–2001), so on, to 10 years of data were used to calculate female survival.

are merely indicators that can be used to focus attention on a herd. Furthermore, these outcomes would be most applicable to herds that number 30 or fewer elk before the decline began. When herd size is 30 or fewer, the window of time to extinction from low adult survival will be much less than, say, a herd that numbers 100 or more. In settings with all the qualifiers described, further action to ascertain agents that are causing female mortality should provide better information to assess whether the decline signals herd extinction.

In the contemporary world, a population extinction can be thought of as something that is lost forever and, therefore, something that should be avoided at all costs, especially when anthropogenic activities can be fingered as hastening the extinction. There are ample examples to justify this view. Even so, extinction is an ecological and evolutionary process involved in metapopulation dynamics and also shapes populations, communities, and ecosystems (Levins, 1969; MacArthur & Wilson, 2001). Instead of focusing on how to prevent extinction in extant herds, an alternative could be to consider repopulation after extinction.

Since the demise of the Boyes herd, repopulation could occur from the females that travel to Boyes Meadows to birth young and that remain with offspring in the meadow until autumn and then leave. This is what I call the birth young possibility. Two other possibilities are nearby herds that expand home ranges to include Boyes Meadows, and colonization by an entirely new group of females.

The birth young possibility seems a remote prospect. To repopulate a herd that remains in Boyes Meadows year-round would require successive years in which, initially, one to two females birthed young that were then recruited. The recruited young that were females would then remain with their mothers and also birth in Boyes Meadows when they reach sexual maturity. If this continued, the group might arrive at a size where these animals decide to remain in Boyes Meadows to exploit forage because the surrounding forest contains little feed. The issue seems to be herd growth starting from one and two females. Solitary elk or a pair of elk that associate together do seem to have increased risk of mortality (Fryxell et al., 2008). The possibility of these few elk being able to recruit young to obtain a group size where mortality risks are less is probably far-flung.

The nearest herds to Boyes Meadows are probably not going to expand their home ranges to include this meadow. Davison and Gold Bluff herds are the closest but occupy meadows that are separated by at least 3 km of forest. Moreover, these two herds are either arriving at equilibrium with their food supply or appear to have already established a *K* carrying capacity (next chapter). Herds near *K* carrying capacity are presumably not faced with elevated lethal risks from predators or a collapse in the food supply. As such, elk are not motivated to venture into unfamiliar areas to seek out new forage habitat (Weckerly, 1999).

Repopulation of Boyes Meadows might happen through colonization. Reliable documentation of elk colonization that then proceeds to abundant populations indicates that the initial number of colonists were 5 to 20 elk (Bender and Piasecke, 2010; McCorquodale et al., 1988). What could prompt that number of colonists is possibly a disturbance (Sousa, 1984). If a forest fire, for example, creates substantial forage habitat near Boyes Meadows, elk might then encounter Boyes Meadows, colonize it, and remain.

Regardless of if and how Boyes Meadows might be repopulated, one thing is acutely apparent. We are witnessing if and when Boyes Meadows will be repopulated by females, juveniles, and subadult males that remain year-round. This unintended experiment is ongoing.

INTERPRETATION

The extinction of the Boyes herd was unexpected. These elk were not hunted, the forage in the meadow presumably was enriched by periodic burning, and the herd had existed for nearly 80 years. There appears to have been time for the herd to reach an equilibrium with its food supply in spite of the irruption that might have occurred in the 1940s and 1950s. Documentation on abundance of the Boyes herd since its likely colonization in 1937 was possible because the herd was involved in numerous studies (Franklin et al., 1975; Gogan and Barrett, 1987; Harper et al., 1967; Weckerly, 1999; Weckerly et al., 2001). The herd gradually declined from 1998 until I no longer detected any females, juveniles, and subadult males in my January population surveys after 2010. The change from a herd size of four in 2010 to none thereafter indicates that Allee effects were likely. The Allee effects might have been from elevated mortality associated with small group size, or these herd members might have moved on and fused with other herds. The latter possibility seems plausible because some females used Boyes Meadows to birth young after 2010. Familiarity with birth sites might be a factor prompting females to return to the same site year after year (Wolf et al., 2009).

The demise of the herd seems to be associated with the Bypass stretch of U.S. Highway 101. The Bypass opened in 1992 and provided meadow foraging habitat that was not there before construction of the Bypass. Seemingly, herd abundance should have increased with the creation of new forage habitat. I started collecting data on spatial patterns of the Boyes herd in the winter of 1995–1996. The Boyes herd primarily used Boyes Meadows and the Bypass. From locations of carcasses and apparent mortalities (i.e., no carcasses were found but there was a drop in group size), I estimated that elk deaths were three times more likely along the Bypass than in Boyes Meadows. Although I could not identify what killed every elk along the Bypass, the likely suspects were vehicle collisions and poaching. Along stretches of the Bypass there are median barriers and a grade. These highway features might restrict elk mobility to evade a vehicle and the driver of a vehicle that encountered an elk probably took longer to slow down in the downslope lanes. Furthermore, this stretch of Highway 101 is relatively remote. Poachers that killed elk along the Bypass might have a low risk of getting caught.

Despite what led to the demise of the Boyes herd, there should be signs in elk demography that associate with the extinction. Female survival was lower in the Boyes herd compared to the Davison herd across years when both herds were declining. Given that the Boyes herd went extinct, it is no surprise that females had low survival. Also associated with the demise of the Boyes herd was greater process variance in female survival and juvenile:female ratios in the decline years. The greater process variance in the Boyes herd was mostly from demographic variance, the stochastic events that affect an individual elk's ability to survive and recruit young. Greater environmental variance in the Boyes herd did not occur and might be considered a surprise. Presumably, the Boyes herd had more extrinsic factors than the Davison herd because of the presumed Bypass agents. There can, however, be greater environmental heterogeneity that does not translate into increased environmental variance (Kendall, 1998). Elk might be able to buffer against increased environmental variance through behavior, physiology, and the herd being far from *K* carrying capacity.

Monitoring the Boyes herd up to 13 years before its demise provided an opportunity to assess how quickly demographic variables might show signs that could result in an extinction. Nipping a problem in the bud seems to always have the greatest chance of working. Analyses that assess what demographic variables affect population growth the most have repeatedly shown that female survival has the greatest influence on population persistence in large herbivores. Patterns that were evident in female survival after 10 years of decline were present with as little as 4 years of decline. Adult female survival lower than 0.80 and a CV of 0.18 or more *might* be indicative of a herd going extinct when the herd size is 30 or fewer. The low adult female survival rate and increased variance should, perhaps, be followed up by more intense herd monitoring to determine what is killing elk. That way a more informed decision can be made about whether remedial action is needed to arrest the herd decline before extinction becomes inevitable.

What is intriguing is why the Boyes herd continued to use the Bypass with its elevated risk of mortality. After all, Roosevelt elk can identify risks from hunting and poaching and adjust movement and habitat use to mitigate those risks (Cole, Anthony, & Pope, 1997; Cole, Pope, & Anthony, 2004). I suspect there are at least two facets to elk having difficulty perceiving risk along the Bypass. One facet might have to do with the challenge that ungulates seemingly have gauging speed and distance of oncoming vehicles (Blackwell, Seamans, & T.L. DeVault, 2014; Meisingset, Loe, Brekkum, & Mysterud, 2014). That challenge, coupled with reduced mobility that elk face from median barriers and the longer time that it probably takes down-slope vehicles to slow down, might greatly challenge elk to recognize and evade these risks. Also, elk in the Boyes herd were habituated to people. Habituated elk, consequently, might also have

FIGURE 5.8a, 5.8b. Photographs of similar views of a stretch of the Bypass showing plant successional changes from meadow to forest. The upper photograph was taken on January 9, 2005, and the lower photograph was taken on January 11, 2014.

difficulty distinguishing vehicles carrying people that wanted to observe them, avoid them, did not detect them, or poach them. The Bypass might constitute an ecological trap to elk (Dywernchuk and Boag, 1972). Elk use the Bypass because of the foraging opportunities that are near the 101, but have difficulty perceiving risks from anthropogenic activities (Meisingset et al., 2014).

Given that elk herds are within 3 km of Boyes Meadows, repopulation of the meadow through colonization seems plausible. Since 2011 Boyes Meadows has been inhabited solely by adult males in January. In early February 2015, two to four males were probably poached along the Bypass. These males had inhabited Boyes Meadows in the previous January when I conducted field work. Lethal risks appear to persist in this area. Whether risks from poaching or vehicle collisions will exist if females, juveniles, and subadult males recolonize the meadow is unknown. The risk might depend on the frequency that a new herd uses the Bypass. If Bypass use is frequent, then mortality might be high and the colonizing elk might not persist. Boyes Meadows might then be classified as a sink habitat because the Bypass is probably an ecological trap (Van Horne, 1983). Elk that disperse to Boyes Meadows do not survive long enough to reproduce and juveniles produced from mothers that inhabit this meadow do not survive to reproduce. On the other hand, if colonists of Boyes Meadows use the Bypass infrequently, then it seems plausible that a new herd can grow, establish a *K* carrying capacity, and persist.

The Bypass stretch of the 101 does seem to be a shadow to elk (see chapter quote). Elk that use this stretch of the 101 might benefit from a reduction in speed limit or sensors that detect elk and flash to warn motorists to slow down and be cautious. I have heard that in the 1970s patrols at all hours of the night alleviated poaching risks to elk. Who knows: maybe succession of plant communities from meadow to redwood-conifer forest will reduce forage habitat along stretches of the Bypass to the extent that elk will not frequent the Bypass and the shadow will disappear (Fig. 5.8).

6

Population Dynamics and *K* Carrying Capacity of Remaining Herds

Efficient management necessitates an understanding of how the managed population works.

— Caughley, "Wildlife Management and the Dynamics of Ungulate Populations"

Bald Hills, Gold Bluffs, Levee, and SOC (the herd that inhabited an area including the old South Operations Center of Redwood National Park) were the remaining herds (Fig. 1.5, 1.6). I viewed the population dynamics of these herds as "normal" for large herbivores. Realizing that normal is in the eyes of the beholder, I considered the population dynamics as unsurprising because the herds did not irrupt, nor did they expire by factors associated with anthropogenic activities. The population process that seems to dominate herd dynamics in the RNSP, density dependence, should be operating and therefore *K* carrying capacity ought to be estimable. Moreover, given the seasonal phenology of the food supply (Fig. 0.5) and when population surveys were conducted, a 2-year lag should be apparent. Fat and protein reserves, which affect whether or not females conceive and recruit young, are presumably deposited in late winter and early spring prior to the autumn mating season. Additionally, population surveys to estimate the number of mouths consuming late winter and early spring food supplies are conducted even earlier, in January. The end result is a 2-year lag (see Chapter 7). Findings from the irruptive Davison herd were consistent with a 2-year lag to density-dependent juvenile recruitment. If population growth rate also has a 2-year lag with abundance, then it should be detectable from a prediction of the standard-logistic model, with a twist. Instead of a standard 1-year lag with intrinsic rate of growth [$r = \ln(N_{t}+_{1}/N_{t})$], there should be a 2-year lag [$\ln(N_{t+2}/N_{t})/2$].

A formulation of the GSS model, on the other hand, treats *K* carrying capacity as a stationary distribution (Dennis et al., 2006). In a straightforward view, *K* carrying capacity tells you the ceiling on the abundance of an elk herd that can persist. For old-timers like myself, *K* was the holy grail that could rarely be estimated from data collected in the field. To estimate *K* requires a long time series. My data from elk in the RNSP indicates 19–20 years. Not only is a long time series required, but estimating the signal (process variance) through the noise (observer error) is also required. An abundance time series with much noise and weak signal will tend to give biased estimates of *K*. It was only within the past 10 years or so that old-timers like myself have had to expand their horizons and learn new techniques. The herd abundance data I collected over the years provided the opportunity to estimate signal, noise, and *K* of multiple herds. Estimates of *K* from multiple herds would then be useful to examining how carrying capacity influences population dynamics (Saether et al., 2008; Street et al., 2015).

According to the standard-logistic model, if there is variation in *K*, then the strength of density dependence might vary across populations (Lande, Engen, & Saether, 2003; Saether et al., 2008). The maximum rate of intrinsic population growth (r_{max}) should be dictated by life history attributes of the species: age of first reproduction, litter size, frequency of reproduction, and survival rates (Street et al., 2015). Landscapes are heterogeneous, and as such *K* carrying capacity will be dictated by the food supply. In a simple view, the food supply is the product of the forage biomass per hectare and hectares of area occupied by the herd. Among populations of a species, life history attributes dictating r_{max} should be constant but *K* will probably vary. The outcome is that populations with lower *K* should have stronger density dependence because the slope of the relationship between *r* and *K* will be steeper than a population that has a higher *K* (Fig. 6.1). The highest *r* on the graph, which corresponds to an abundance of zero, is r_{max}. In other words, the estimate where *r* is unencumbered by food limitations.

The strength of density dependence has utility for assessing the tendencies that might happen to population dynamics. Relative to herds with lower *K*, herds with higher *K* might react more weakly to changes in abundance whether abundance is close to as well as farther from *K*. Irruptions, therefore, should be more likely in herds with higher *K*. Also, herds with a higher *K* might display larger fluctuations in abundance than herds with small *K*.

Another way that variation in *K* carrying capacity should influence population dynamics is in the endogenous and exogenous factors that affect

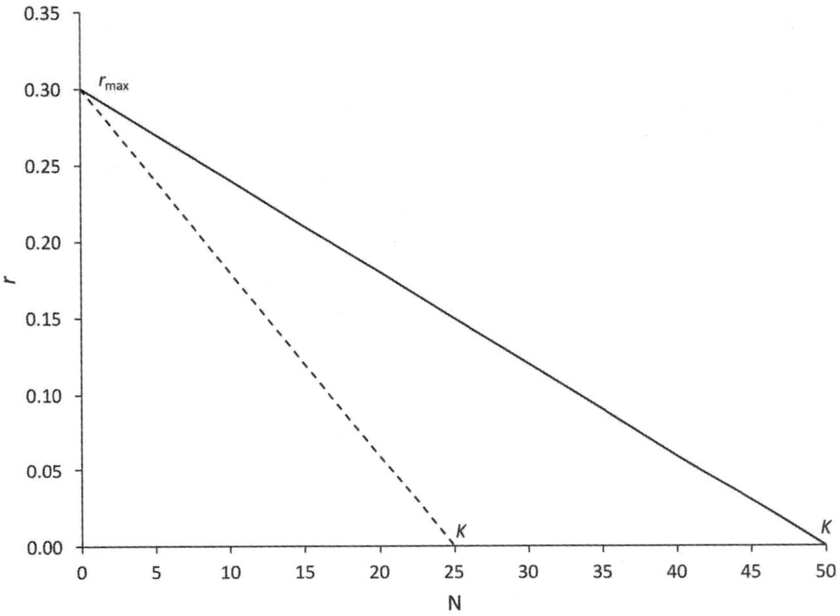

FIGURE 6.1. Conceptual graph showing change in strength of density dependence for two hypothetical herds that have the same maximum intrinsic rate of increase (r_{max}) but differ in K carrying capacity.

population change. As described in Chapter 5, endogenous factors are labeled demographic and exogenous are labeled environmental. Changes in abundance (N) from each of these factors are measured by demographic (σ_D^2) and environmental (σ_E^2) variances. Demographic variance is the difference in survival and reproduction across individuals that seemingly occurs by random chance. A variety of factors that are labeled stochastic affect individuals such that individual variation in demography can be thought of as resulting from random chance. Environmental variance originates from abiotic and biotic factors that impact survival and reproduction of all individuals in the population. These two variances are the components of process variance:

$$\sigma^2 = \sigma_E^2 + \frac{\sigma_D^2}{N} \qquad 6.1$$

(Lande et al., 2003). Process variance is a reflection of the biological factors that drive change in abundance. The implication of Formula 6.1 is that herds with lower K will have greater σ_D^2 because small abundances are more affected by stochastic events that affect survival and reproduction of individuals. A real-world example was the large change from year to year in juvenile:female

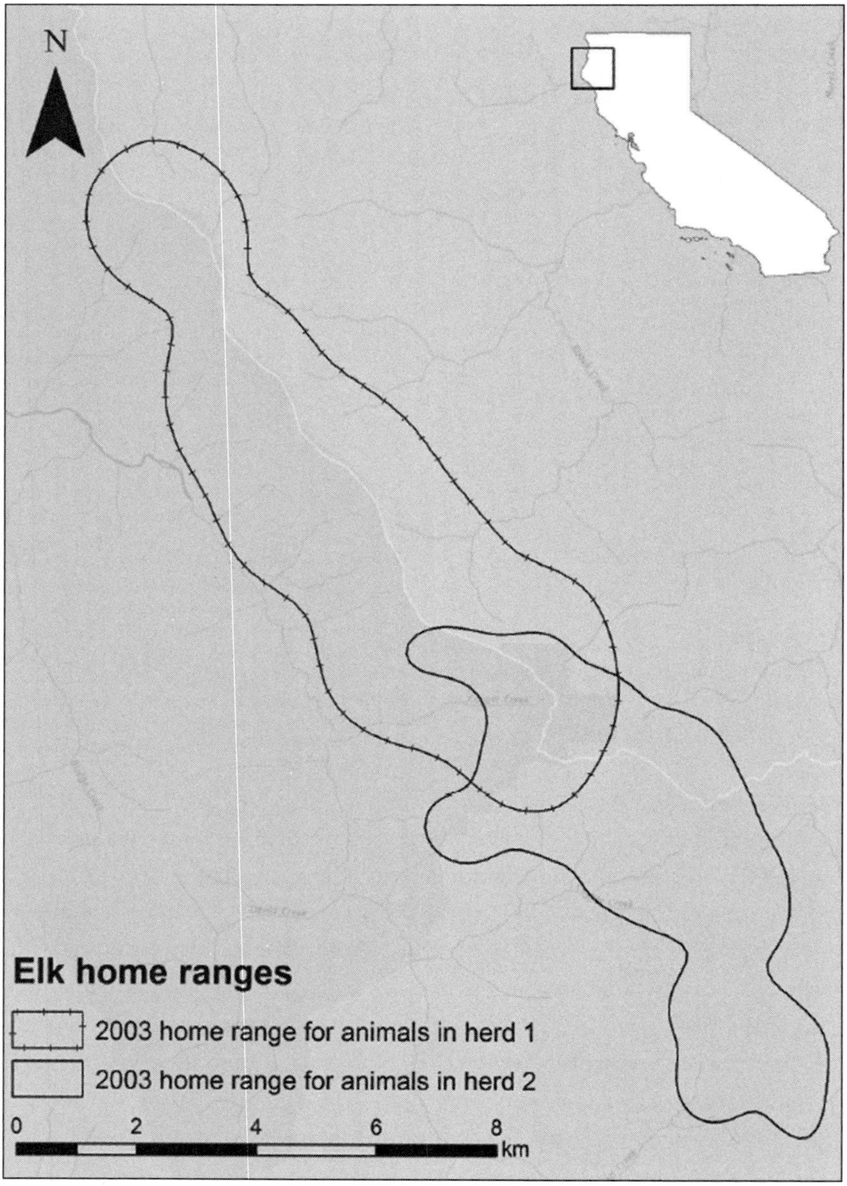

FIGURE 6.2. Annual home ranges of the northern (Herd 1) and southern (Herd 2) herds in Bald Hills in 2003 (Kolbe & Weckerly, 2015). Home ranges were estimated with the fixed kernel estimator and an ad hoc smoothing parameter. The northern herd had one collared female and the southern herd had three collared females with highly correlated movements (Kolbe & Weckerly, 2015).

ratios when the size of the Boyes herd was small compared to when it was larger (Fig. 5.1). Across herds there should be an inverse relationship between process variance and *K* carrying capacity (Saether et al., 2008). When there is a smaller *K*, extinction is more likely because σ_D^2 comprises a greater part of σ^2. Of course, what is removed from consideration in Formula 6.1 is noise or temporal changes in abundance that can arise from imprecisions in the population surveys. The imprecisions across counts from survey methodologies are collectively labeled observation or observer error (τ^2). Imprecisions occur because the places viewed during a survey might not include the places where elk are located, or some individuals might not be sighted because they are obstructed by vegetation, terrain, and other elk. Human foibles might also result in missing animals that should be detected or detecting elk but misclassifying them to the wrong age–sex category.

The following presentation of individual herd dynamics is in alphabetical order: Bald Hills, Gold Bluffs, and finally Levee and SOC. After presenting individual herd findings, an interpretation among herds is then offered.

BALD HILLS

The complex of meadows in the Bald Hills might actually contain two herds and not one (Fig. 6.2). In 2003 six females were fitted with GPS transmitters (Kolbe and Weckerly, 2015). The transmitters were programmed to take locations every 5 hours and batteries powered the transmitters for 1 year. Four of the females were captured and fitted with transmitters on the south end of Bald Hills and the other two females were from the north end of Bald Hills. Transmitters were recovered from four of the six females and data downloaded into spreadsheet files. The mechanism that released the neck collar from the elk failed in the two GPS transmitters that were not recovered. The data recovered from the four transmitters indicate that the movement of three of the females seemed to be as one animal because the locations were clustered together (Kolbe and Weckerly, 2015). Viewing Figure 6.2 it appears that the locations of these three females overlapped little with locations from the remaining elk. On a monthly basis, the average degree of overlap in home ranges was 11 percent. These findings indicated that there were two herds in the Bald Hills (Kolbe and Weckerly, 2015).

In spite of these findings from the GPS data collected in 2003, I assume that there was only one herd across the entire time series. I made the assumption for two reasons. In 2005 location data on herds were not collected, and in some of the other years I was unsure that there were two distinct herds.

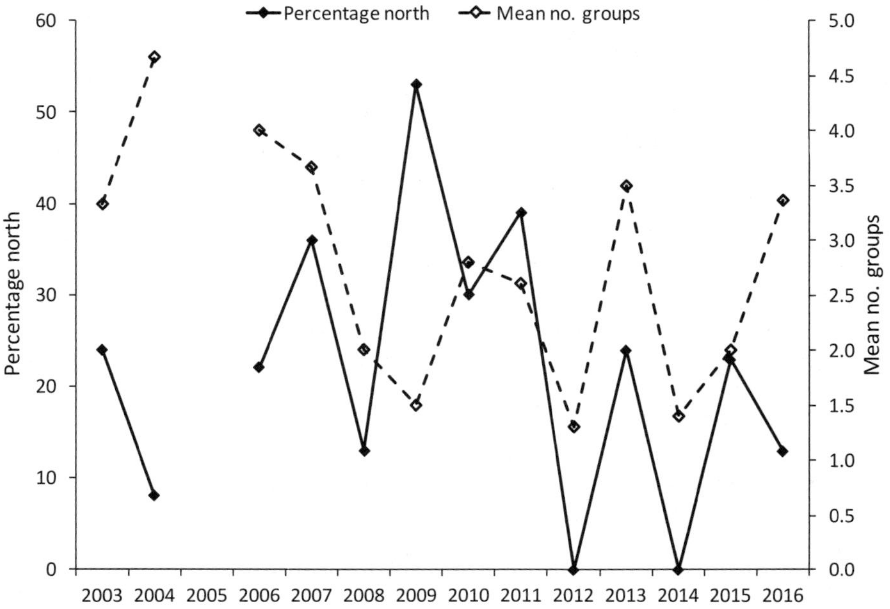

FIGURE 6.3. Line chart showing the mean number of female groups encountered during surveys in the Bald Hills (2003–2004, 2005–2016) and the percentage of groups that were sighted north of the locations of Females 1, 2, and 4 in Figure 6.2. Data on distribution of groups encountered were not collected in 2005. Female groups were predominantly comprised of females, juveniles, and subadult males.

Consequently, I could not place the group encountered during surveys as belonging to the north or south herds. To show this I used some of the population survey data collected in 2003 and 2004, and between 2006 and 2015. In each of these years, I calculated the mean number of female groups encountered during surveys and the percentage of groups that were located to the north of the locations of Females 1, 2, and 4 (Fig. 6.3). The mean was calculated from groups encountered regardless of whether the view of the group was obstructed or unobstructed. In 2012 and 2014, in particular, the mean number of groups was 1.3 and 1.4 (medians in both years were 1.0), respectively. In those 2 years, all the groups encountered were to the south of where Female 3 was located in 2003. The boundary between the north and south herds might actually have shifted farther to the south. Yet, I did not have GPS data from the two herds beyond 2003. For the sake of simplicity, I viewed the females, juveniles, and subadult males encountered in the Bald Hills across the time series of data as composing one herd.

It took 28 years for the Bald Hills herd to increase from 4 to more than 200 elk (Fig. 2.9). Since 2004 herd abundance has fluctuated between about 180 and 276 and ostensibly indicates that the population is near *K* carrying capacity. As a reminder, the count of 180 in 2016 might have been a severe case of observer error (Chapter 2). Perhaps I simply missed substantially more elk than what I probably did in other years. Eyeballing Figure 2.9, the abundance estimates and 95 percent credible intervals obtained from the GSS model summarized the dynamics of the Bald Hills herd. Estimates of the parameters a, c, σ, and τ are needed to estimate *K* (Dennis et al., 2006). These estimates also appear precise considering the credible intervals of these parameters (Table 6.1). Density dependence affected the dynamics of the herd because c was less than 1.0. The *K* carrying capacity was projected to be 235, which is somewhat lower than most abundances between 2004 and 2015. It would seem that *K* should be closer to the highest abundance observed (276) across the years from 2004 to 2015 when abundances appeared to plateau. The estimate of *K* in the Bald Hills time series might be a result of the noticeable amount of observer error. Process variance as a percentage of the sum of process variance and variance from observer error (Φ_1) was noticeably less than for the other two herds listed in Table 6.1.

Unlike the areas occupied by the other herds, prescribed fire was a disturbance constantly applied in the Bald Hills. Burning is widely reported to benefit large herbivores through increasing the food supply (Kolbe, 2016). Prescribed fire was a land management tool used by the RNSP to reduce conifer and coastal redwood tree establishment in meadows. Without burning, conifers and redwoods readily colonize meadows in the Bald Hills, grow rapidly, and displace herbaceous vegetation (Fritschle, 2008). Burning, by the way, was a practice used by Native Americans before Europeans arrived in the mid-1850s, and the RNSP instituted the prescribed fire program to restore the cultural practice. For the years of data (2002–2015) used by Kolbe (2016), the hectares of meadows burned each year was not consistent. The mean percentage of meadow hectares burned across these years was 26 percent but ranged from a low of 2 percent to as much as 81 percent of the meadows being burned. Prescribed fire was a density-independent disturbance because it was not conducted to explicitly manage elk forage habitat and herds.

The Bald Hills herd might also be susceptible to density-independent climatic influences. Elk in the Bald Hills inhabited an area that was farther inland and up to 900 m higher in elevation than the Prairie Creek and lower Redwood Creek drainages. As such, temperature extremes and precipitation were probably greater in the Bald Hills (Starns et al., 2015).

TABLE 6.1. Parameter estimates of GSS model for three herds obtained from Bayesian analysis. The parameters were *a*—rate of population growth in the absence of density dependence, *c*—strength of density dependence, σ—process standard deviation, and τ—standard deviation of observer error. From these parameters, or squares of these parameters, carrying capacity (*K*) and its standard deviation, Φ_1 and Φ_2, were calculated (Dennis et al. 2006). Percent variation in the time series due to process variance is Φ_1 and Φ_2, the percent of variation in the stationary distribution due to process variance.

HERD	PARAMETER	MEAN VALUE	STANDARD DEVIATION	2.5TH PERCENTILE	MEDIAN	97.5TH PERCENTILE
Bald Hills	*a*	0.623	0.090	0.465	0.615	0.822
	c	0.885	0.019	0.843	0.887	0.919
	σ	0.091	0.030	0.039	0.090	0.155
	τ	0.062	0.026	0.032	0.056	0.124
	K	235	45	—	—	—
	Φ_1	72	—	—	—	—
	Φ_2	92	—	—	—	—
Gold Bluffs	*a*	0.893	0.535	0.000**	0.889	1.951
	c	0.700	0.169	0.366	0.701	1.031*
	σ	0.233	0.045	0.162	0.228	0.335
	τ	0.052	0.028	0.032	0.043	0.123
	K	21	6	—	—	—
	Φ_1	97	—	—	—	—
	Φ_2	98	—	—	—	—
Levee–SOC	*a*	1.550	1.035	0.000+	1.466	3.994
	c	0.601	0.268	−0.031	0.622	1.058*
	σ	0.116	0.035	0.038	0.118	0.183
	τ	0.061	0.034	0.032	0.047	0.158
	K	49	7	—	—	—
	Φ_1	86	—	—	—	—
	Φ_2	91	—	—	—	—

* Indicates a 97.5th percentile for *c* that is greater than 1.0 suggesting density independence. In a long-lived species like elk, I expect *c* to be less than 1.0. Therefore, the 95th percentile should be less than 1.0 (Seavy et al. 2009). The 95th percentile was 0.975 for Gold Bluffs and 0.960 for Levee–SOC herds, indicating that density dependence affected the population dynamics of these two herds.

** The 2.5th percentile was a negative value, which is unrealistic for *a*. Consequently, I report zero for those two values.

Nonetheless, prescribed fire, temperature, and precipitation had no detectable influence on juvenile recruitment in the Bald Hills herd (Kolbe, 2016). Similar to the Davison herd, juvenile recruitment appeared to be driven by density dependence. Prescribed fire, in particular, should have influenced juvenile recruitment because the abundance of the Bald Hills herd was close to and often above *K* carrying capacity in many of the years between 2004 and 2015. Burning should have increased the food supply, resulting in an increase in juvenile recruitment. It appears that not enough of the Bald Hills Meadows was burned from year to year to increase the food supply to the extent needed

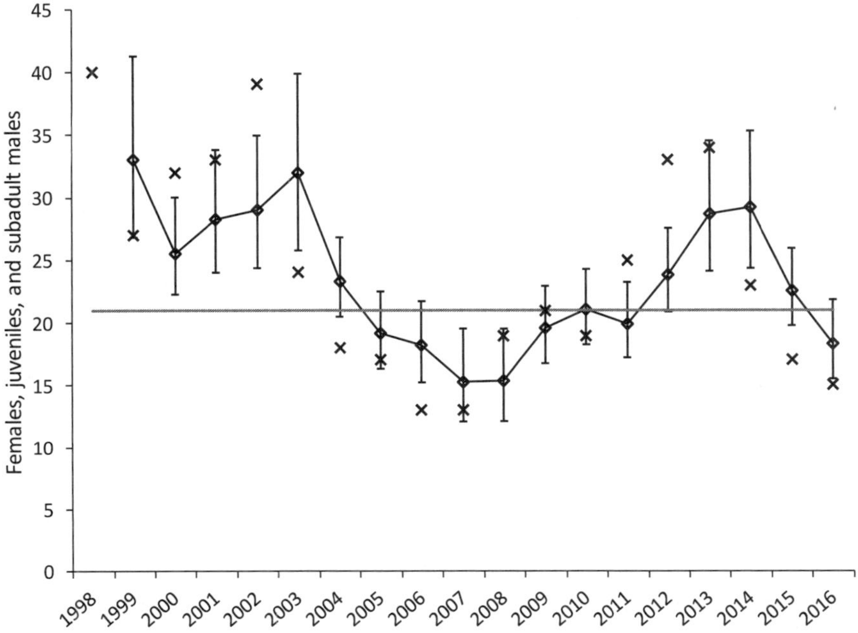

FIGURE 6.4. Time series from 1998 to 2016 of abundance estimates of the Gold Bluffs herd (X) along with estimates, and 95 percent credible intervals, from a GSS model (diamonds and lines). The gray horizontal line is the estimated *K* carrying capacity.

for a population response. Density dependence appears to be the predominant driver of the Bald Hills herd dynamics, similar to most herds at lower elevations near the coast.

GOLD BLUFFS

The time series of the Gold Bluffs herd began in 1998 and ostensibly displayed a cyclical pattern (Fig. 6.4). Also, herd abundance was rather low, fluctuating between 13 and 40. The cyclical population dynamics displayed by the herd seemed to be captured by the GSS model. Unlike the estimates for the Bald Hills model, however, estimates of some of the GSS parameters were imprecise, the estimate of *a* in particular. The credible intervals of *a* ranged from 0 to 1.951. The credible intervals of *c* were also wide. Yet, density dependence influenced the Gold Bluffs herd dynamics because the 95th percentile of *c* was less than 1.0. The estimate of *K* carrying capacity was 21. Parameters with more precise estimates were σ and τ. Observer error appeared to have little influence

TABLE 6.2. Findings on relationships between herd abundance and intrinsic rate of population growth calculated with 1-year ($\ln[N_{t+1}/N_t]$) and 2-year lags ($\ln[N_{t+2}/N_t]/2$). For each herd, the simple linear regression coefficients and standard errors are reported for the stronger relationship as indicated by the r^2. The time series was 19 years long for each herd.

HERD	LAG	r^2	INTERCEPT	SE	SLOPE	SE
Gold Bluffs	One	0.19				
	Two	0.34	0.305	0.131	−0.014	0.005
Levee–SOC	One	0.16				
	Two	0.36	0.420	0.144	−0.009	0.003

on the temporal variation of herd abundances as 97 percent of the variation in the time series was estimated to be from process variance and the remaining 3 percent from observer error (Φ_1, Table 6.1).

The precision of the estimate of parameter *a* is alarming. Because *a* is needed to estimate *K* in the GSS model (Dennis et al., 2006), the estimate of *K* might also be incorrect or biased. Consequently, I also estimated *K* in another way.

The abundance estimates from the count data seemed to be little affected by observer error. The Gold Bluffs elk and all others in the lower Redwood Creek and Prairie Creek drainages were habituated to people (Bowyer, 1981; Franklin & Lieb, 1979; Richey & Wheeler, 2001). My analysis of detection probabilities (Chapter 2) was from elk inhabiting the Prairie Creek drainage. That analysis also indicated little observer error. Findings from analyses of detection probabilities and the GSS model concur: observer error had little influence on abundance estimates of herds habituated to people. Using the herd size data, I then regressed abundance versus intrinsic rate of population growth with a 1-year and 2-year lag (Table 6.2).

The 2-year lag had the strongest relationship. A scatterplot of the data and regression indicated that *K* was 22 (Fig. 6.5), quite close to the estimate from the GSS model. The *K* carrying capacity estimate of 21 obtained from the GSS model appears to be plausible.

LEVEE AND SOC HERDS

The Levee and SOC herds occupied areas surrounding the town of Orick and in the town itself (Fig. 1.5). In 2015 the Levee and SOC herds fused (Fig. 3.10). Coinciding with the fusion of the two herds was that gates to a 20-ha cattle pasture just south of Orick were opened in autumn 2014. I observed elk in the

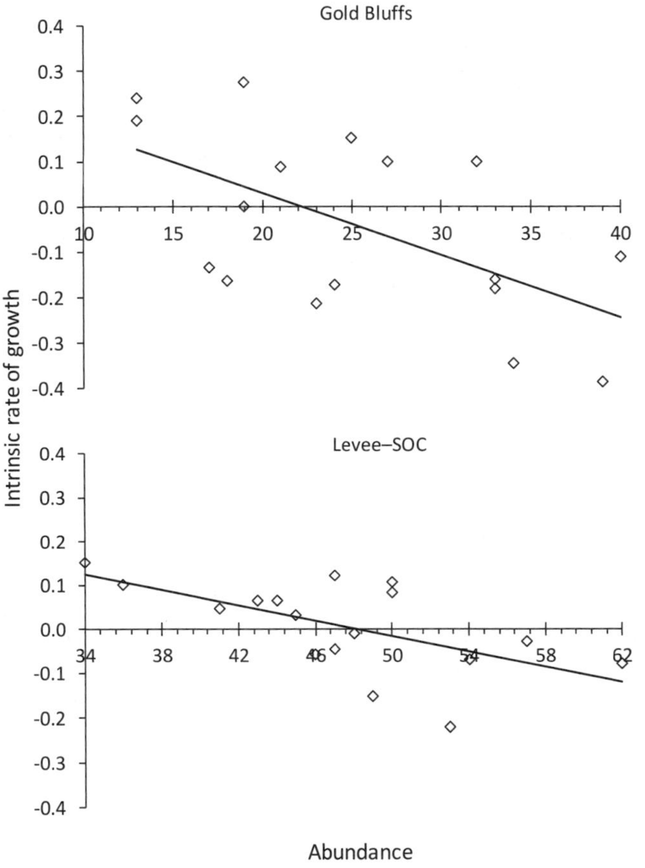

FIGURE 6.5. Scatterplots with regressions between abundance and intrinsic rate of population growth lagged 2 years ($\ln[N_{t+2}/N_t]/2$) for the Gold Bluffs and Levee–SOC herds.

pasture for the first time in January 2015. The opening of the gates might have occurred to make elk accessible to hunting on privately owned pasture. Prior to autumn 2014, elk had not been sighted in this pasture because the southern boundary of this pasture had a 2-meter-tall elk resistant fence (Hofstra, 1997). The fence was built in 1991 to prevent elk in the RNSP from grazing privately owned pastures to the south and east of Orick.

The Levee herd had a marker female: there was one female that had a large slit in her left ear. In 2014 the Levee herd numbered 30 and the SOC herd had 17 elk. In 2015 the Levee herd increased to 59 and I observed the marker female with the elk in the cattle pasture. The 1-year jump in size of the Levee

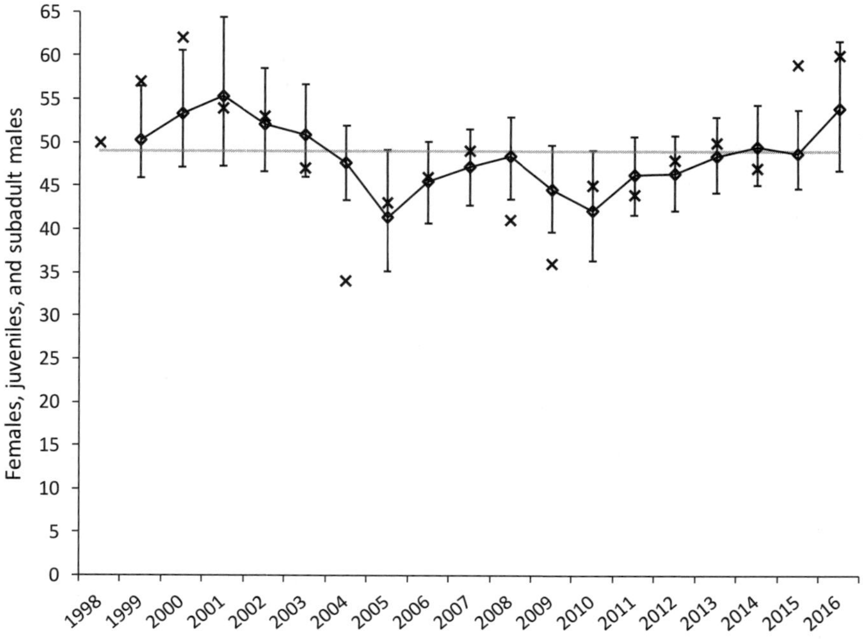

FIGURE 6.6. Time series from 1998 to 2016 of abundance estimates of the Levee–SOC herd (X) along with estimates, and 95 percent credible intervals, from a GSS model (diamonds and lines). The gray horizontal line is the estimated *K* carrying capacity.

herd (intrinsic rate of increase = 0.69) was beyond what was possible from juvenile recruitment alone (see Chapter 4). But juvenile recruitment can explain the increase if the 2014 counts of both herds are summed (ln[59/47] = 0.23). Furthermore, I did not detect the SOC herd by itself in 2015. It seems plausible that the Levee and SOC herds fused to exploit the new food source in the cattle pasture. Further glue to maintain fusion of the two herds might have come from the lethal risks posed from hunting that were possible in the autumns of 2014 and 2015.

The dilemma then was whether to analyze the data of the two herds separately or together (i.e., sum the counts in each year). I chose to analyze the data together because the counts of the two herds across 1998 to 2016 were not independent because of the 2 years when the two herds fused into one. The combined herds were then called the Levee–SOC herd.

Between 1998 and 2015 the Levee–SOC herd fluctuated between 34 and 62 elk (Fig. 6.6). Similar to the Gold Bluffs herd, the GSS model appeared to summarize the population dynamics but some of the parameter estimates were suspicious (Table 6.1). In particular, *a* appeared to be a high estimate again, and imprecise. In spite of the imprecision of the estimated *c*, density dependence appeared to affect the dynamics of the herd. *K* carrying capacity was estimated to be 49. Process variance accounted for 86 percent of the temporal variation in the time series.

For reasons articulated with analysis of the Gold Bluffs herd, I also estimated *K* from regressions between abundance and *r* lagged 1 and 2 years (Table 6.2). Similar to findings from the Gold Bluffs herd, the relationship was stronger for the 2-year lag. Eyeballing the scatterplot of the data and regression indicates that *K* is 48 (Fig. 6.5). This estimate of *K* is close to the estimate from the GSS model (Table 6.1). Again, the *K* carrying capacity estimate of 49 obtained from the GSS model appears to be plausible.

K AND FORAGE ABUNDANCE

A more fundamental way to assess whether *K* carrying capacity estimates resemble the actual *K* carrying capacity is to estimate the food supply. When there is more food, there should be a greater *K* carrying capacity. The bulk of the food supply for elk herds in the RNSP should be meadow forage. For the Bald Hills and Davison herds, I have estimates of the area of meadows occupied by these herds and monthly NDVI data from those meadows across 23 years (Starns et al., 2015). The Bald Hills herd occupied about 1,000 ha of meadow. The Davison herd (Chapter 4) occupied 51 ha of meadow. The areal densities of *K* (*K*/ha of meadow) were 0.235 (235/1000) for the Bald Hills herd and 1.06 (projected *K* of 54) for the Davison herd. The much greater areal density of *K* for the Davison herd suggests that forage biomass per hectare of meadow should be greater than in the Bald Hills. To examine this possibility, I analyzed the monthly NDVI values in Bald Hills and Davison Meadows across 23 years of data. With the exception of December, the NDVI values in the Davison Meadows were greater than in the Bald Hills Meadows (Fig. 6.7). The NDVI data indicate that the lower areal density of *K* in the Bald Hills was related to the herd having to range over a large area to obtain forage. The Davison herd had more feed per hectare of meadow and did not have to range as widely as the Bald Hills herd.

INTERPRETATION

There was wide variation in *K* carrying capacity among the three herds. The magnitude of the difference between the largest and smallest *K* was (235–21) 11.2 times. The potential is there that the magnitude might be affected by lumping two herds into one in the Bald Hills. Lumping two herds into one should

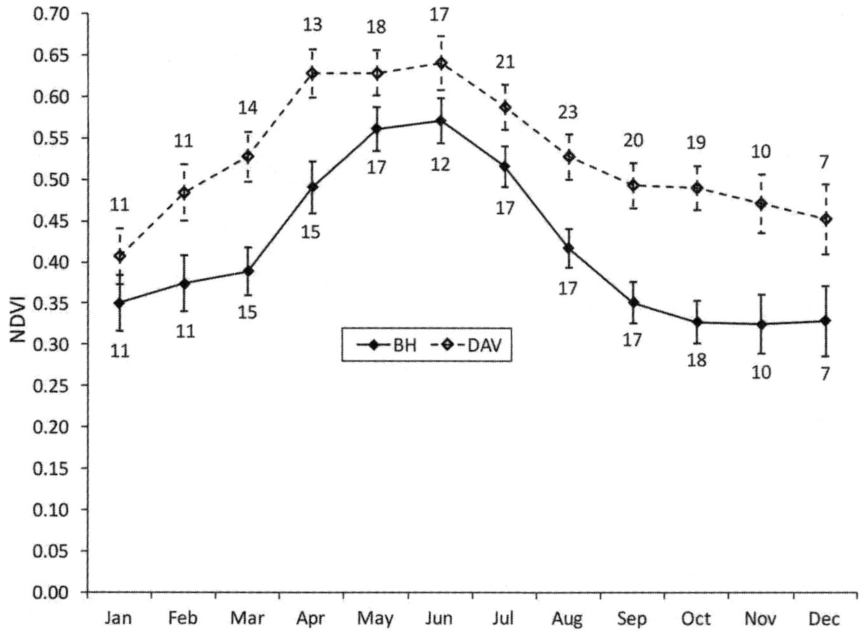

FIGURE 6.7. Line plot of monthly mean NDVI, and 95 percent CI, for the Bald Hills and Davison Meadows complexes for data gathered from 1989 to 2011. Numbers of year are by CI. A two-factor analysis of variance indicated that there was an interaction between month and meadow complex ($F = 3.03$, $df = 11, 327$, $P < 0.001$).

result in a larger estimate of *K*. Be that as it may, for moose (*Alces alces*) populations in the province of Ontario (Street et al., 2015) and populations of various waterfowl species in the prairie pothole region (Saether et al., 2008), the magnitude was much less (2–3.2). The area considered for the moose study was thousands of square kilometers and the waterfowl study was across 8–10 degrees of latitude. The area encompassing the three elk herds was only about 50 km^2. Because the aerial extent of the moose and waterfowl studies was so much greater, it is somewhat surprising that *K* varied as much as it did in my elk study.

Part of an explanation for why *K* varied so much across the elk herds probably has to do with how populations were delineated. In my study, populations were defined by the area occupied by herds. The geographic boundaries of these herds appear to be self-contained. Herd growth was not coupled to growth of any other herd. In the moose and waterfowl studies, populations were delineated by how population surveys were conducted. Growth of one population, therefore, might have been coupled to growth of others. In the moose study, a population was defined as a game management unit; in the waterfowl study, a population was delineated by transects flown during aerial surveys.

Lumping data for two herds and calling them one herd in the Bald Hills and for the Levee and SOC herds was a practical necessity. In the Bald Hills, I could not distinguish two herds in some years and in 1 year the location of encountered herds during population surveys was not collected. I could not tell if the north and south herds fused or remained separate. When the Levee and SOC herds fused in 2015, two herds that previously had independent population dynamics (Julian et al., 2013) suddenly were the same herd. But because I could not distinguish each individual in herds, I was unable to estimate the influence of individual demography and movement on population dynamics within and between herds (Lee, 2015). When herds fuse and individuals of each herd cannot be identified, estimating demography and population dynamics from lumped data is probably inevitable.

Lumping or pooling data masks variability (Bhaskar and Weckerly, 2016). Consequently, lumping the time series of two herds into one will do the same. Lumping abundance data also will result in an increased estimate of *K*. A higher *K* also suggests less process variance because of lower demographic variance (Formula 6.1). Process variance was lowest for the Bald Hills herd ($0.091^2 = 0.0082$), intermediate for the Levee–SOC herd ($0.116^2 = 0.0135$), and greatest for the Gold Bluffs herd ($0.233^2 = 0.054$). The respective *K*s of these herds were 235, 49, and 21 (Table 6.1). Although those are just three data points, *K* carrying capacity appeared to be inversely associated with process variance, which is a finding that was similar to the findings from a much larger data set of various waterfowl species (Saether et al., 2008). Lumping abundances of two elk herds into one should result in reduced estimates of process variance, particularly when *K* is small—say, 20–30.

An estimate of process variance is also needed to estimate variance of *K* or the stationary distribution (Dennis et al., 2006; Lande et al., 2003). If *K* is inversely related to process variance, then the stationary distribution might also be less variable, relative to *K*, when *K* is large. The coefficient of variation (CV) of *K* for the three herds does not appear to follow this expectation (Table 6.1). For the Bald Hills herd, the CV is 0.19, for the Levee–SOC herd it is 0.14, and for the Gold Bluffs herd it is 0.29. The standard deviations of the *K*s appear to be indicative of their stationary distributions because the percentages of the standard deviations that are process variance (Φ_2) are high (Table 6.1). For the Levee–SOC herd in particular, I thought the standard deviation of *K* would be great, relative to *K*, because *K* had to change across the time series when the

20-ha cattle pasture became available. The precision of that *K* estimate might have been relatively high because when each herd was separate, density dependence was pervasive and abundances fluctuated around *K*. From 1998 to 2014, years before the herds fused, the dynamics of each herd was impacted by density dependence; *K* carrying capacity was also estimable for each herd (Levee: 95th percentile of $c = 0.951$, $K = 32$; SOC: 0.77, $K = 14$). When the cattle pasture became available, the increase in available forage habitat was enough for both herds to fuse. Yet, forage limitations were not substantially alleviated such that density-dependent feedbacks with the new food supply lessened substantially.

The parameter estimates of the GSS models were most precise for the Bald Hills herd. Both the Gold Bluffs and Levee–SOC herds had time series that did not begin well below *K*, a feature that is evident in the Bald Hills time series. More-precise parameter estimates are usually obtained when a time series begins well below *K* (Dennis and Taber, 1994; McCullough, 1979). Because of my discomfort with some of the parameter estimates from the GSS model for the Levee–SOC and the Gold Bluffs herds, I also estimated *K* via a more conventional approach. I used this approach because the count data for these two herds should be little affected by observer error. The estimates of *K* from the regressions were similar to the GSS model, reducing my anxiety and increasing my confidence in the estimates of *K* from the GSS model. Other data that indicated *K* estimates might be informative were the NDVI data. The NDVI was a surrogate of forage biomass and the NDVI data were consistent with the higher areal densities of *K* for the Davison herd compared to the Bald Hills herd. By the way, the areal density of K for the Bald Hills herd was probably an overestimate. That estimate did not consider the meadows that were likely to be used by the Bald Hills herd when they used Stover's Ranch (Fig. 6.2).

It was nice to see that the GSS model detected density dependence and provided estimates of *K* similar to what I observed from expectations of the standard-logistic model. The Gompertz model appears to be most applicable to species like waterfowl that tend to have lower survival rates and higher fecundity than a long-lived species such as elk (Dennis et al., 2006). In long-lived ungulates, the standard-logistic model is probably more applicable. To my knowledge, the only comparison between fit of the Gompertz and standard-logistic model to the same elk time series was Taber and Gogan's (2002) analysis of Yellowstone elk. Fit of one model to the abundance time series was not substantially better than the other, albeit neither model accounted for observer error. Indeed, more work is needed to parameterize standard logistic models such that observer error and process error can be estimated so that a robust comparison can be made between the two models.

When I analyzed data for this chapter, I had hoped to be able to assess the strength of density dependence. Unfortunately, I am unable to do so with the existing data sets. For the GSS model, the credible intervals of parameter *c*, the estimate of the strength of density dependence, overlapped considerably across the three herds. It is evident from the standard errors and coefficients of the two herd regressions in Table 6.2 that 95 percent CI of these coefficients would overlap as well. The strength of density dependence might indeed vary according to *K* as presented in Figure 6.1. But if that is the case, then a longer time series is needed to be able to obtain parameter estimates that are likely to be more precise. With such data, it will then be possible to detect statistically significant differences in the parameter estimates that would distinguish changes in the strength of density dependence among herds.

It is also feasible to expect that the strength of density dependence might not weaken with larger *K*. Expecting a herd's *K* to remain constant across the decades it takes to attain an estimate from time series data might be unrealistic. Climatic patterns can and do vary at the scale of decades and centuries, and landscape composition of habitats often change because of human activities and disturbances such as fire and flood. Another difficulty with estimating the strength of density dependence is when a time series of abundances that span the spectrum from low to *K* is lacking. This was an issue with some of my herd data where most of my abundance estimates were clustered around *K*. Indeed, more data are needed to ascertain whether the strength of density dependence is inversely related to *K* among populations of an ungulate species.

7

Conservation and Management

There is a need for greater awareness of the role that serendipity plays in detecting ecological surprises by ensuring that at least some of the scientific endeavor in ecology is curiosity driven rather than tightly constrained by reporting against narrowly focused milestones and producing results in line with preordained deliverables.

— Lindenmayer et al., "Improved Probability of Detection of Ecological 'Surprises'"

POPULATION ECOLOGY AND CONSERVATION

One of the fundamental decisions was seemingly mundane: how to define a population. Yet, how I defined a population had profound implications for what I could examine in the data and the patterns I detected. I decided to go with a collection of socially bonded females, juveniles, and subadult males—a herd. The next decision was one of hierarchical scale for analysis of population data. Should a population consist of components such as herds in an area that was often in close proximity and sometimes fused, or should each herd be a separate population? Going with a population hierarchy or metapopulation is useful such as when individuals disperse between habitat islands (Bleich, Wehausen, & Holl, 1990). In such a setting, population abundance would be the sum of herd abundances. Furthermore, a time series of those summed population abundances should probably show less variation than the variation at the level of the herd. If I had gone with metapopulation, then I could have delineated four populations: Bald Hills, Boyes, Gold Bluffs, and the lower Redwood Creek and Prairie Creek drainages. In the lower Redwood Creek and Prairie Creek drainages, lumping the Davison, Levee, and SOC herd counts

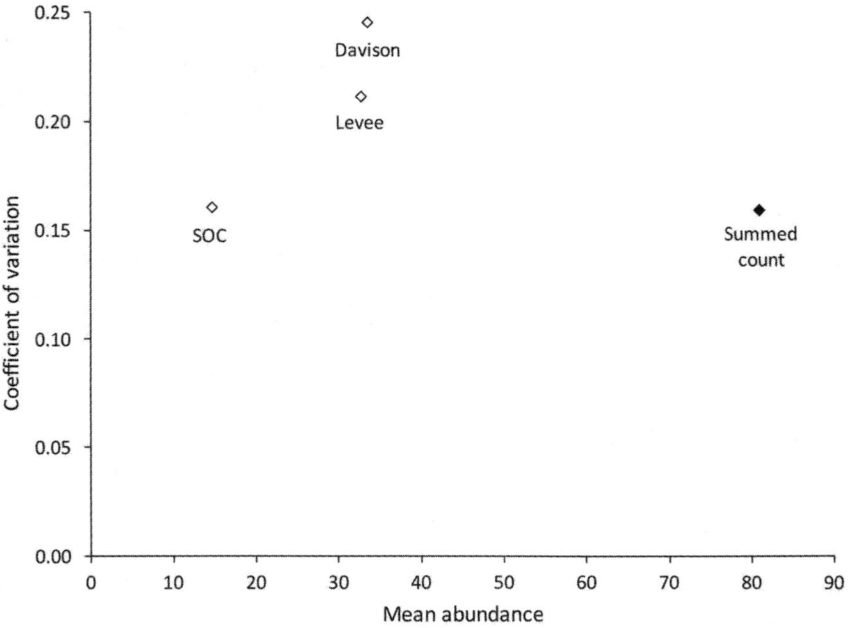

FIGURE 7.1. Scatterplot showing means and coefficients of variation across 1998–2014 for each of the Davison, Levee, and SOC herds, as well as when the counts of these herds were summed. Across these years, the herds were often in close proximity to one another and two of the herds (Davison and Levee) were documented to fuse for 3 days in 1 year.

into one, across the years when there were three herds, did not necessarily result in reduced temporal variance (Fig. 7.1). The SOC herd, which had the lowest abundance, displayed about the same temporal variation in abundances as was found when the counts of all three herds were summed. This result is not surprising, given that all available evidence suggested little mixing of individuals among herds such that growth rates were uncoupled between herds (Franklin & Lieb, 1979; Julian et al., 2013; Weckerly, 1999). Also to be remembered, the fusion of the Levee and SOC herds and the Davison irruption might have gone unnoticed if these herds were considered one population. Across small geographic areas for Roosevelt elk, herd dynamics can behave very differently.

Because how I defined a population was fundamental to everything I did, it would make sense that I formed the definition in the beginning. Paradoxically, I did not form the definition in the beginning (Weckerly, 1996, 1998, 2007; Weckerly et al., 2001; Weckerly et al., 2004). I was sold on viewing the herd as the population unit only after I had enough years of data to show independent herd dynamics (Julian et al., 2013). I might not have had the time to gather evidence to examine how to define populations and the hierarchical scale of analysis if there had been contractual agreements that dictated population summaries early in the study. Given my ignorance at the time, I probably would have set up arbitrary populations that did not consider which females associated and which did not. Although the shoestring budget imposed many limitations, there were liberties.

Defining a population as a herd most likely allowed me to encounter unexpected findings. As the unexpected findings were certainly not anticipated or viewed as likely when I began, they can be regarded as ecological surprises (Lindenmayer et al., 2010). Uncovering ecological surprises usually has conservation and management implications, and my study was no exception. Irruption by the Davison herd might have eroded resilience, and the demise of the Boyes herd eroded redundancy and identified a potential sink habitat. Uncovering ecological surprises appears to be associated with four conditions. One, the objectives of the study did not include uncovering ecological surprises. A surprise, by definition, cannot be anticipated. Two, data are collected through long-term studies. At the very least, long-term studies increase the potential to detect a finding that was not anticipated or thought to be possible. I would not have been able to detect the irruption of the Davison herd or the extinction of the Boyes herd if I had stopped the study after 5 or 10 years. Three, comparisons are possible by conducting parallel studies so that the unexpected can be uncovered. In my study, multiple herds were tracked. When I labeled the Davison herd dynamics an irruption, it called into question whether irruptions were a common occurrence for colonizing herds in the RNSP, particularly since the herd that colonized Boyes Meadows in the late 1930s might also have gone through an irruption. That the Bald Hills herd did not go through an irruption suggests it is not a given for every colonizing herd in the RNSP. Four, disturbances seem to be associated with ecological surprises. The building of the Bypass and the change in ownership of the Davison Meadows from private to public property were sudden changes that probably began the process of shifting communities away from the state that existed before the changes (Sousa, 1984). Both of those disturbances were associated with the irruption and extinction surprises. Another condition that also might be associated with detecting ecological surprises is minimizing or removing lumping or pooling at the level of the basic units by which data are collected. The basic unit of data collection for me was the herd. In essence, an ecological surprise is a finding that manifests as something beyond what is perceived as the usual variation present in all ecological systems. If data are collected at the basic unit, then variability will not be masked and surprises, when they occur, can be unmasked. In my study, the inability to determine whether there were two herds in the Bald Hills potentially masked variability and, perhaps, precluded me from detecting additional surprises.

Herd distribution was associated with forage distribution and abundance and, apparently, social behavior. Meadow partitioning between herds occurs and might have been maintained by social fences. As a consequence, determining the spatial boundaries of herds might not be straightforward. In other words, measuring and describing the distribution and abundance of food at the landscape scale will not necessarily reveal the number of herds in that landscape. Understanding how meadow partitioning can manifest and be maintained, then, has relevance to determining the number of herds in a landscape and the setting under which a change in the number of herds might happen, information that is vital to assessing redundancy (see the Introduction).

Social fences are associated with animals that are familiar with a spatially and temporally predictable food supply. Yet, herds face food limitations (Kolbe & Weckerly, 2015; Mysterud et al., 2011). The manifestation of a social fence is from social interactions between unfamiliar animals of different herds in a zone of overlap. These interactions interfere with food acquisition exacerbating the already food-limited setting. For a social fence to come down, then, requires a new food source to alleviate food limitations. Increased food intake then offsets costs from aggression while unfamiliar animals are becoming acquainted with one another. Multiple, small-sized herds might become one large-sized herd such as what happened in the fusion of the Levee and SOC herds.

Across the 20 years of this study, a fusion of two herds occurred. What about in the other direction? When might a single herd split into two or more herds? Food limitations might suddenly become more severe if, for example, a pasture that is grazed by a herd suddenly becomes unavailable when real fences are erected or repaired and the gates are closed. One large herd might split into two herds that install social fences. Costs of aggression from interactions with numerous animals in a large herd exacerbates the keen, food-limited setting. In a large herd, the extent of familiarity with all individuals should also be more heterogeneous than among members of a small herd. Social fences are maintained by social interactions but seem to require food limitations to manifest.

The driver of population dynamics was density dependence. Density-dependent herd dynamics should be prevalent when nonmigratory elk herds with spatial boundaries inhabit a landscape that has a seasonal, predictable food supply associated with a mild, wet climate (Bowyer et al., 2014). In such a setting, negative feedbacks should develop between herd size and the food supply. Indeed, my data indicate that was the case. Even when abundances were high and near *K* carrying capacity, which is when density-independent factors should be easiest to detect, density-dependent processes had a pervasive influence on population dynamics (Kolbe, 2016). The only instances where density dependence was not detected was in the Boyes herd as it was descending into extinction and when the Davison herd was in the decline phase of the irruption. Because both were unexpected and vexed me to no end while they were occurring, I initially was reluctant to label them as anomalies. Perhaps density-independent climatic factors might have been influential. Approximately 400 km to the south of the RNSP at the Point Reyes National Seashore, density-independent precipitation interacted with abundance to affect elk population dynamics (Cobb, 2010; Howell et al., 2002). In the end, however, I observed little to indicate density-independent precipitation influenced population dynamics in the RNSP. Climatic patterns at Point Reyes are somewhat drier and warmer than in the RNSP. In a mild, wet climate the plant growing season is probably longer and the biomass of forage after the growing season might also be greater than at Point Reyes (Conroy, 1987; Johnson and Cushman, 2007). For climatic variables to dramatically reduce food supplies in the RNSP might require extreme conditions such as a severe drought that lasts multiple years (Kolbe, 2016). Across my 20-year study no such drought occurred.

Critical to assessing the influence of environmental factors on herd dynamics is an understanding of the time it takes for herds to respond to abiotic and biotic influences (Caughley, 1976). The anticipated feature to density-dependent herd dynamics was the lagged response between herd abundance, juvenile recruitment, and population growth. The consistent finding across herds was a 2-year lag. I think there are two reasons for the 2-year lag. One reason is because of the mismatch between the windows of time when nutritious forage is available and the season of parturition is followed by females provisioning young with milk. Throughout the year, nutritious forage is likely to be most abundant during the period of rapid plant growth around March and April (see the Introduction). The nutritional value of forage is high during plant growth (Crawley, 1983). Cell-soluble material in plants is abundant and much of it is readily digested relative to cell wall material. Much of the digestible material in the cell wall requires time-consuming fermentation (Barboza et al., 2009). Yet, most births occur from mid-May through June after the period of rapid plant growth. In late winter and early spring females might have to replenish protein and fat stores to see them through the energy demands of the latter part of gestation and lactation. In summer and much of autumn there appears to be forage but not an abundance of nutritious forage. Much of the forage is dormant or senesced and leaching of plant nutrients seems likely from fog and autumn rains (Tukey, 1970). Large herbivores such as elk are capital breeders (Taillon, Barboza, & Cote, 2013). Animals store nutritional reserves (capital) to see themselves and their young through times when nutritious forage is scarce. Maybe the nutritional reserves acquired during the growing season also have a large influence on whether females can conceive young in the mating season in late summer and early autumn.

The other reason for the 2-year time lag in demographic and population response probably had to do with the timing of population surveys in relation to the plant growing season and reproductive events. Surveys to estimate abundance and juvenile recruitment were conducted in January but conception and the onset of gestation occurred in late summer and early autumn, followed by the birth of offspring in May or June. Assuming a mismatch affects female fecundity and recruitment of young, then the 2-year lag works as follows. The herd abundance that captures the number of mouths foraging during the plant-growing season is the count conducted the prior January (lag zero). In January 1 year later (lag one) females are pregnant but it is not until 1 year later in the following January (lag two) that I measured juvenile recruitment. As a hypothetical alternative, if I had conducted population surveys in July shortly after parturition, the population lag most likely would have been 1 year. The herd abundance in July after the plant-growing season is the abundance coupled to female condition that should dictate the possibility that a female conceives in the following mating season. Recruitment would then be measured from population surveys conducted 1 year later in July (lag one). I detected a 2-year lag because of the mismatch between reproductive events and the food supply in relation to when my population surveys were conducted.

Whether Roosevelt elk herds in other parts of their geographic range display a lagged population response has not been examined. It might be possible to test the hypothesis with herds to the north of the RNSP. In the Olympic peninsula of Washington State, the timing of reproductive events appears to be similar to the RNSP (Schwartz & Mitchell, 1945). Why there is similarity in timing of reproductive events between these two places that are about 750 km apart is unclear. It might be because of the large body size of Roosevelt elk relative to other subspecies of elk. Birth mass and milk production seem to scale to the 0.75 power of female body mass (Bhaskar and Weckerly, 2016). The implication is that the mass-specific demands of reproductive events might be greater in smaller females than in large female Roosevelt elk. Consequently, the mass-specific demands from gestation and lactation might not be great enough to force Roosevelt elk females to time birth and lactation to annual peaks in nutritious forage. As long as females can store protein and fat capital, and there is forage available during lactation, then the timing of births and the mating season might follow a common pattern for Roosevelt elk. Mating occurs in late summer and early autumn and females give birth in late spring and early summer. What does differ between the Olympic Peninsula and the RNSP is the seasonal timing of nutritious forage (Gogan & Barrett, 1987; Leslie, 1982). On the Olympic Peninsula, the availability of nutritious forage and the peak in food abundance is in the summer months of June and July. As a result, I do not expect a lagged population response for Olympic elk herds because the timing in availability of nutritious forage coincides with the timing of birth and lactation.

When a herd has discrete spatial boundaries and herd dynamics are driven by density dependence, then the herd should arrive at an equilibrium with its food supply or *K* carrying capacity. The Bald Hills, Gold Bluffs, and Levee–SOC herds appear to be in equilibrium with their food supplies, and the Davison herd appears to be arriving at an equilibrium with its food supply. It is useful to know the time it takes a colonizing herd to arrive at equilibrium with its food supplies. If a herd is well below *K*, then it is difficult to assess the drivers of population dynamics and resilience cannot be evaluated. Surprises also seem to be more likely when a herd is not near *K*. The time it takes to reach *K* depends on the number of colonizing elk, how fast the herd grows, and the abundance of forage in the occupied area. If there are few colonizing elk, and *K* is high, it simply takes longer for the herd to grow to a size at which negative feedbacks with the food supply can be detected through abundance and juvenile:female ratios. Yet, in the real world the time it takes for a herd to reach equilibrium can be complex. When the Levee–SOC herd fused and colonized a newly available cattle pasture, a new *K* appeared to establish quickly. Apparently both herds were in equilibrium with their food supply before they fused and the new cattle pasture probably provided an increase in the food supply but not a hefty boost. There appear to have been around four elk that colonized the Bald Hills in 1978. About 25 years later, in 2004, the herd appeared to reach its *K* of 235. The colonizing group was 1.7 percent of *K*. The Davison herd is projected to reach *K* in 2019, 28 years after Davison Meadows became elk habitat. Assuming 10 elk colonized Davison meadow, probably an underestimate, and that *K* is 54; the colonizing animals were about 18 percent of *K*.

The irruption of the Davison herd and the change in plant species composition and, presumably the food supply, is one explanation for why the Davison herd appears to be taking a long time to reach equilibrium, even though the Davison herd had more colonizing animals and a lower *K* than the Bald Hills herd. Insofar as my data allow, it can take up to 20–30 years for a colonizing elk herd to reach *K* carrying capacity in the RNSP.

In the heterogeneity of the real world, it is intriguing that I was able to detect *K* carrying capacity across a 20-year study. Herd spatial boundaries are likely to change, which should prompt a change in food supplies and a new *K*. The spatial boundaries of the Levee and SOC herds, as an example, did change. I was able to capture that change in 2015 and 2016 when these herds fused and colonized a cattle pasture. The spatial boundaries of the Levee herd also varied from 2007 to 2012, a time when they used the southern end of the Davison Meadows. What is critical to *K*, however, is not necessarily a change to herd spatial boundaries, but rather a change in food supplies. It is possible that the spatial boundaries for a herd change across time but there is not enough of a change in the food supply to precipitate changes in *K* that can be detected from abundance time series.

K carrying capacity is related to the conservation concepts of redundancy and resiliency. Redundancy refers to the number of herds in a landscape and the connections between herd dynamics. The more herds with population dynamics independent of one another, the greater the redundancy and the more likely it is that the species will persist when one or a few herds go extinct (den Boer, 1981). From 1997 to 2010 I witnessed the extirpation of the Boyes herd. Yet, Roosevelt elk persisted in the southern part of the RNSP because there were four other herds. These four herds also had dynamics that appeared to be uncoupled from the Boyes herd and each other.

Resiliency refers to how resistant a population is to disturbances or factors that negatively affect abundance. Estimating resilience is an active scientific topic with many developments over the past 20 years (e.g., Duarte, 2015). I assess resilience in a crude manner. The greater the *K* carrying capacity, the more resilient the herd. Based on this criterion, the Bald Hills herd is clearly the most resilient. I have three reasons for why I did not conduct an involved modeling exercise such as a population viability analysis (PVA) to estimate resiliency. First, findings from a PVA might suggest that more forage habitat is needed to ensure persistence. Because forage habitat would most likely mean removing trees, the signature feature of the RNSP, this option is unrealistic. Second, when PVAs are conducted there is a tendency to view the PVA as good enough for conservation and management. What I mean is the PVA is used as justification for ceasing data collection or stopping any motivation for starting to collect data. After all, neither money nor time for empirical research grows on trees. Third, predicting the future is difficult. The insightfulness of a PVA is tied to the environmental setting and ecological processes fleshed out for the modeling exercise. If, for example, environmental settings in the future change beyond what was anticipated, then the utility of the PVA wanes. As a consequence of these three reasons I resort to crude, qualitative statements about resilience. Although my conclusions about resilience are crude, they are based on empirical data. In my view, there is no substitute for empirical data that provides reliable information about distribution and abundance in perpetuity.

Estimates of process variance are helpful to connect the dots among resiliency, redundancy, and *K*. Process variance is a quantitative estimate that is the sum of environmental and demographic variances (Formula 6.1). Process variance captures the extrinsic and endogenous factors that cause herd abundance to change. From the four data points that I gleaned from my studies and two other data points extracted from the literature, there appears to be an inverse, threshold relationship between *K* and process variance (Fig. 7.2). The nonlinearity appears to be because of the large influence of demographic variance on process variance when abundance and *K* is small, as suggested by Formula 6.1. When *K* for the number of females 1 year or older in a herd is less than about 25, process variance is up to one order of magnitude greater than when *K* is higher. Above a *K* of 25 there is not a dramatic decline in process variance. Relative to *K*, herd abundance is less variable and probably less susceptible to chance events precipitating herd extinction. Hypothetically, then, there is the potential of a trade-off between redundancy and resiliency. A landscape with one herd with a high *K* will be resilient but the species will lack redundancy. No matter how high *K* might be, if that herd goes extinct, then the species also is extirpated. Conversely, if a landscape has numerous herds with *K*s below 25, then there is redundancy. Yet, each herd has low resilience. Figure 7.2 demonstrates that to maximize elk persistence, a landscape should have redundancy with each herd having a *K* above 25 so that process variance is relatively low and resiliency is high. Of course, there is no possibility of directly maximizing elk persistence because the landscapes in the RNSP are not managed for elk. The reported relationship between redundancy and resiliency might be applicable to thinking about elk conservation among the *K*s in a particular landscape (Saether et al., 2008). I had herds with *K*s that ranged widely. Therefore, herds with smaller *K*s and abundances should be more susceptible to

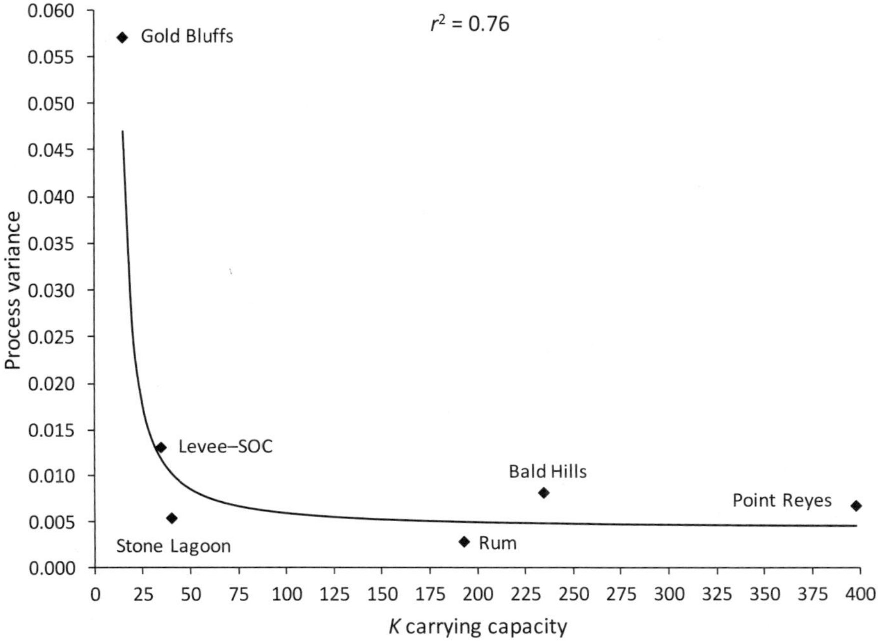

FIGURE 7.2. Scatterplot with regression of the relationship between *K* carrying capacity and process variance of six herds—four from my studies and two others extracted from the literature. All herds experienced little to no hunting. *K* carrying capacity was estimated from the GSS model for abundance of females that were 1 year and older except for Bald Hills, where abundance was for females, juveniles, and subadult males. The relationship estimated was natural log of process variance = −5.494 + 36.539*1/*K* ($F = 13.01$, $df = 1, 4$, $P = 0.023$). The Point Reyes data were obtained from Cobb (2010); the data from the Island of Rum, Scotland, were extracted from Coulson et al. (2004). Number of years of count data for estimating *K* for each herd were Gold Bluffs: 19, Levee–SOC: 19, Stone Lagoon: 16, Rum: 20, Bald Hills: 19, Point Reyes: 20.

extinction, as I observed when the Boyes herd went extinct. The demise of the Boyes herd had to also reduce redundancy of elk in the RNSP.

Abundance, and the largest sustainable abundance possible in a herd, signified by *K* carrying capacity receives considerable attention in discussions of resilience. Another facet to resilience are the processes driving population dynamics (Staples et al., 2005). For a long-lived species like elk, the process of attention is density dependence. I consider density dependence as critical to resilience because the process is required to establish *K* carrying capacity. If a herd has established *K*, then it is possible for the herd to be sustainable and persist. A byproduct of a herd that has established *K* is high female survival and some degree of resilience. The case in point is the Boyes herd that went extinct. Abundance of the herd during my study was 30 or fewer; there was no hint of density dependence in the dynamics of abundance and female survival was low. In a long-lived species like elk, resilience should perhaps consider abundance, density dependence, and *K* carrying capacity.

MANAGEMENT

Wildlife management was originally defined by Professor Aldo Leopold (1933, 3) as "the art of making land produce sustained annual crops of wild game for recreational uses." At that time, decimation of wildlife was rampant, overgrazing from domestic livestock was prevalent, and erosion of topsoil was not a pressing issue until the soil was mostly gone. I could go on. As a consequence, not much study had gone into how to restore degraded land so the process was trial and error or an art: Let's try this and see what happens. Recreational uses were primarily geared toward hunting at the time. How can hunting opportunities be increased and at the same time not overexploit game populations? In the RNSP, recreational use is not hunting elk, but viewing them. As long as there are elk to view, the land is producing an annual crop. As such there is no pressing need to restore the land to produce an annual crop. Indeed, the need for elk management in the RNSP is limited. Why study elk for management purposes?

Elk management in the RNSP, and other parks for that matter, is limited. Yet, management actions to effect a desired result are possible and do occur. Prescribed burns, mowing of meadows, and other anthropogenic alterations to parklands occur (Kolbe, 2016). Furthermore, Roosevelt elk can be translocated from the RNSP to portions of the historical range where they were extirpated (Logsdon, 1965). Elk are also a charismatic mega-fauna. Because of the keen public interest in elk, studying this species demonstrates stewardship of a public trust. Demonstrating stewardship also provides public education about dynamics of elk distribution and abundance. Attention to public education is becoming a more pressing need because North Americans are essentially urban and, as a result, are less familiar with the workings of wildlife populations and ecological systems. Indeed, stewardship and public education can be a reason for management. There can be a reason to study elk for purposes of conservation and management, and, at the same time, a reason to leave them alone (Caughley and Sinclair, 1994).

Managing efficiently is a common call (Caughley, 1976). To me, efficiency is mostly about finances and the data that are collected and how those data are summarized so that management goals can be assessed. For elk in the RNSP, a long-term study was needed to assess the drivers of distribution and abundance dynamics. A long-term study requires a consistent supply of money. Obtaining consistent funding over decades has the highest probability of success when there are not many dollars needed in the first place, what I call a shoestring budget. Efficient data collection can be described, then, as maximizing data collection relative to the shoestring budget. Efficient data summarization requires an understanding of how populations work and putting that knowledge to work focusing attention on management goals. The overarching goal of this study is assessing redundancy and resiliency.

Obtaining abundance data can drain a budget. Methods for estimating elk abundance are quite expensive when aerial surveys are involved or when elk are marked so that mathematically involved estimators can be used to account for observer error when estimating abundance. In contrast, count data using straightforward, cost-effective survey methods that rely on vehicles and hiking are more cost-effective. RNSP personnel also do not require extensive training to participate in surveys, and the survey methods are readily grasped by students. The hiking is good exercise for most of us, as well. At the same time, there always seems to be a drawback. In my case, I had to forgo counts of total abundance (all age–sex classes) and focus on counting the female segment of the population. That was because female counts do not appear to be dramatically affected by observer error because females used meadows much of the time, which was not necessarily the case with adult males. Without expensive technology like telemetry, I simply did not have the capability to count males or estimate the extent of imperfect detection of males in Bald Hills, Gold Bluffs, and the lower Redwood Creek drainage. If I had decided to include males in population estimates without the use of telemetry, the following saying comes to mind: "A half of a behind plus a half of a behind still equals a half of a behind." The saying, of course, is improved with more colorful language.

Determining herd distribution or actual population boundaries is rarely done. Yet, given the enormous range in *K* carrying capacity uncovered herein, delineating actual population boundaries is a useful endeavor. A convenience about herds in the RNSP was the high degree of social bonding and the relatively small ranging distances of most herds. Consequently, I was not burdened by the expenses of needing fixed-wing aircraft or helicopters to survey large areas. Social bonding among females is not prevalent in all parts of the geographic range of elk (Franklin and Lieb, 1979). In nonmigratory elk, social bonding seems to be affiliated with predictability of seasonal forage and the landscape composition of foraging patches (Jenkins and Starkey, 1984; Street et al., 2013). In places where elk movement is concentrated because forage patches are discrete and interspersed in a landscape matrix with little forage, social bonding among females is evident (Byers, 1997; Weckerly, 1999). Along the west coast, both tule and Roosevelt females display social bonding (Cobb, 2010; Jenkins and Starkey, 1984). Managers planning long-term studies of elk distribution and abundance might be able to use this behavior to develop and use more-cost-effective methods to measure herd distribution.

When designing population surveys to detect socially bonded herds, the survey design will have an element of where to survey. Survey areas must include the areas inhabited by the targeted herds. Yet, survey designs must also be able to accommodate the possibility of herd changes in distribution and imperfect information about herd distribution. Herds adjust spatial boundaries when new food sources become available or when availability to forage patches is lost. The possibility that herds fuse must also be considered. Accommodating temporal changes in herd distributions usually means surveying areas beyond where herds are sighted when a study begins. Critical to obtaining unbiased measurements of herd distribution is to survey forage habitat where initial observations indicate elk are frequently located as well as rarely or never located.

Essential to understanding population dynamics is studying demography (Duarte, 2015; Gaillard et al., 1998). Understanding the roles of inputs (births and immigrations) and outputs (deaths and emigrations) is needed to understand temporal and spatial variation in population abundance. Marking animals is the usual means to collecting demographic data. Yet, measuring all four demographic attributes from marked individuals can also drain a budget. For socially bonded herds where density dependence is the main driver of abundance, year-to-year change in abundance mostly seems to be from juvenile recruitment. Movement of females among established herds appears to be scarce and survival of females 1 year of age and older is high and fluctuates much less than for juveniles during the first half year of life (Griffin et al., 2011). The only time I uncovered some other demographic process influencing herd dynamics was the likely ingress of individuals during the rapid growth stage of the Davison herd irruption. Assuming that ingress into vacant or new habitat is an uncommon occurrence, an efficient use of limited funds would be to focus on measuring juvenile:female ratios. This ratio is probably the demographic variable most likely to inform about changes in abundance. If juvenile:female ratios are measured shortly after the season of parturition, then these ratios are likely to indicate female fecundity, or the proportion of females that give birth. If the ratios are measured when juveniles are half a year old, then juvenile recruitment into the herd is likely to be captured. Survival of juveniles in the last half of the first year of life is much higher and less variable than during the first half-year of life (Harn, 1958; Lieb, 1973). Measuring juvenile:female ratios when juveniles are older than a half year but have not reached the age of 1 year is likely to incur a measurement error. Juveniles might be confused with subadult and adult females, and vice versa, because of large body size due to the additional time for body growth.

Getting back to the particulars of managing elk in the RNSP, it might be beneficial to summarize data at two levels: at the level of the herd and at the level of all herds. All of my findings point to the merits of not masking heterogeneity by lumping herds to understand the drivers of distribution and abundance and to assess redundancy and resiliency. Which bodes for summarizing data at the level of the herd. Nevertheless, RNSP biologists have to make decisions about elk persistence considering all the herds in the RNSP. Wildlife management is not exempt from complexity.

Presently, RNSP biologists summarize time series data annually (e.g., Schmidt, 2015); this is a practice that demonstrates stewardship and should be continued, in my view. It might be instructive to organize data into equilibrium herds, herds where *K* carrying capacity has been estimated, and not-equilibrium herds. Equilibrium herds are Bald Hills, Gold Bluffs, and Levee–SOC herds. A herd not at equilibrium is the Davison herd. Distinguishing herds as such should be useful for conservation matters. Herds not in equilibrium are the ones that can have issues with resiliency and, therefore, might have greater conservation concerns. For this herd categorization to be relevant requires annual population data and periodic updating of *K* carrying capacity. Updating of *K* carrying capacity also is useful to assessing if *K* is changing across time. Perhaps every 5 to 10 years the data should be collated and *K* carrying capacity should be estimated again.

When herd dynamics are uncoupled and asynchronous, and when there is wide variation in *K* and herd sizes, it is likely that the sum of abundances of all herds should display temporal patterns not evident in time series of each herd (Schindler, Armstrong, & Reed, 2015). All females, juveniles, and subadult males in all herds in the southern part of the RNSP can collectively be called

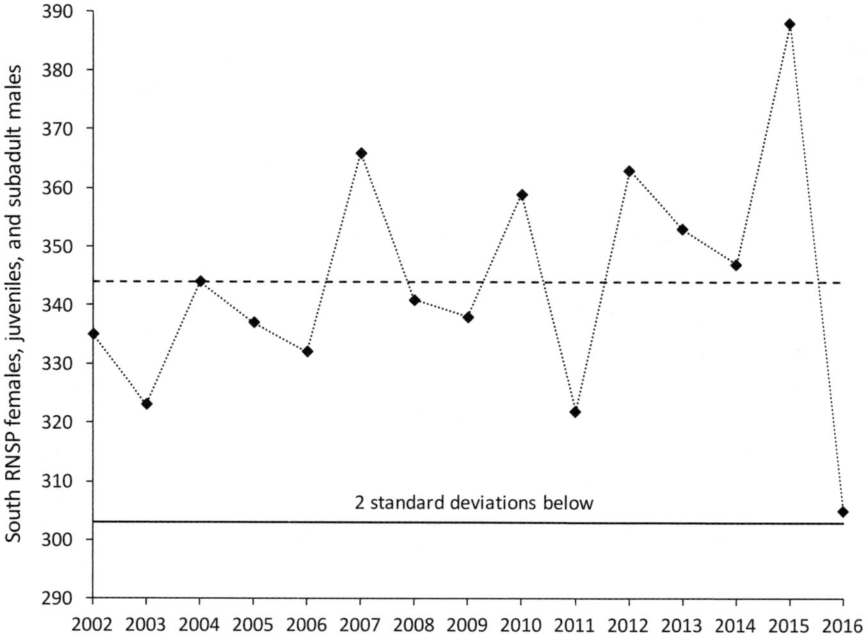

FIGURE 7.3. Line chart showing time series of south RNSP female abundances from 2002 to 2016. South RNSP refers to the summed counts of the five herds: Bald Hills, Boyes, Davison, Gold Bluff, and Levee–SOC. There was not a detectable relationship between year and south RNSP abundance ($Y = 335.5 + 1.007X$, test of slope: $t = 0.81$, $df = 13$, $P = 0.431$). The horizontal dashed line, therefore, is the mean abundance (344). The solid, horizontal line is two standard deviations below the mean.

the south RNSP females. The south RNSP females show a stationary population trend (Fig. 7.3), a pattern quite different from what was detected in each herd across the same years. If RNSP biologists desire an abundance that indicates cause for concern, perhaps a south RNSP abundance that is in the lower 2.5th percentile (about two standard deviations below the mean) might be that criterion. A retrospective illustration of how the criterion can be used is the Boyes herd. That herd declined and went extinct between 1997 and 2011. Across those years, the Boyes herd abundance ranged from 30 to 4. In spite of the decline and extinction of the Boyes herd, there was not a detectable decline in the south RNSP females, in part because the abundance of the Boyes herd was a small part of the south RNSP females. Also, in the years after the Boyes herd went extinct there has not been a noticeable decline in the south RNSP females below the two standard deviations line. Restoring a Boyes herd is not necessarily a pressing issue when considering redundancy and resilience of other herds.

Managing efficiently and effectively might benefit from data summaries at the herd level and, also, at the level of the south RNSP.

The illustration of the lack of influence of the Boyes herd on the south RNSP females brings out the uneven abundance across herds. That unevenness segues to the large role of the most abundant herd, the Bald Hills herd, on south RNSP persistence. Between 2002 and 2016, years when I surveyed all extant herds, the Bald Hills herd averaged 69 percent ($s = 6$) of the abundance of south RNSP females. The high *K* estimated for the Bald Hills herd indicates this herd is the most resilient. Yet, if the Bald Hills herd does decline dramatically in the future then persistence of all south RNSP females will erode. The large *K* of the Bald Hills herd relative to the other herds can be viewed as both an asset and a cause for concern. If there is one herd that requires more attention precisely because it is abundant, it is the Bald Hills herd.

Long-term ecological studies motivated by conservation and management questions are often labeled targeted and surveillance studies (Wintle, Runge, & Bekessy, 2010). Targeted studies are designed to address clearly stated hypotheses. The workings of population dynamics are unclear and so a study is conducted to test competing hypotheses to understand the processes driving population change. Dale McCullough's (1979) study of the George Reserve white-tailed deer herd in Michigan is a classic example. Surveillance studies, on the other hand, are not motivated by the idea of testing specific a priori hypotheses. Instead, attributes of a species or a range of species are measured and the information collected is freely available. The intent is that someone will take these data and analyze them to assess environmental conditions, discover unusual patterns, or provide ecological insights that might have conservation and management implications. A popular example is the North American Breeding Bird Survey (Sauer, Fallon, & Johnson, 2003).

I would not label my long-term study as either type. I think of my study as a stewardship study. To manage and conserve elk, I first needed to understand the drivers of distribution and abundance given the peculiarities of the landscapes. I then took those findings and assessed redundancy and resiliency. This entailed the study of familiar concepts like density dependence, density independence, time lags, and *K* carrying capacity. There were also mundane details to be worked out that had far-reaching implications. Notably, what is the minimum length of time to conduct the study? What is a population and should I use population statistics obtained from herd counts or estimates that accounted for observer error? What had not been examined in much detail was coupling elk social behavior to foraging. This was needed for greater awareness of the dynamics of herd distributions—namely, the possible uncovering of social fences and the mechanism for their origin and maintenance. Along the way, I had the opportunity to uncover some unexpected findings that I viewed as surprises. In a nutshell, identifying the surprises required awareness of the dynamics of distribution and abundance, information that I think is useful for stewardship of a public trust, public education, and, of course, conservation and management of elk in the RNSP.

Bibliography

Ackerman, B. B., L. Kuck, E. H. Merrill, & T. P. Hemker. (1984). Ecological relationships of mule deer, elk, and moose. In *Cooperative wildlife: phosphate study*, edited by L. Kuck, 151–268. Boise, ID: Idaho Department of Fish and Game.

Akçakaya, H. R. (2002). Estimating the variance of survival rates and fecundities. *Animal Conservation* 5, 333–336.

Allen, M. L. (2014). The ecology and behaviour of pumas (*Puma concolor*) in northern California, U.S.A. (Unpublished doctoral dissertation). University of Wellington, Victoria, New Zealand.

Arnold, T. W. (2010). Uninformative parameters and model selection using Akaike's Information Criterion. *Journal of Wildlife Management* 74, 1175–1178.

Arthur, R. L., R. J. Hudson, N. T. Donkar, & J. R. King. (2009). Foraging rates of wapiti on alfalfa/bromegrass pastures. *Small Ruminant Research* 82, 46–52.

Asmus, J., & F. W. Weckerly. (2011). Evaluating precision of cementum annuli analysis for aging mule deer from southern California. *Journal of Wildlife Management* 75, 1194–1199.

Atwood, T. C., E. M. Gese, & K. E. Kunkel. (2009). Spatial partitioning of predation risk in a multiple predator–multiple prey system. *Journal of Wildlife Management* 73, 876–884.

Barboza, P. S., K. L. Parker, & I. D. Hume. (2009). *Integrative wildilfe nutrition.* New York: Springer.

Barnes, E. (1925a). A few Roosevelt elk still exist in Del Norte County. *California Fish and Game* 11, 142.

Barnes, E. P. (1925b). Elk in Del Norte County. *California Fish and Game* 11, 90.

Bender, L. C., & J. R. Piasecke. (2010). Population demographics and dynamics of colonizing elk in a desert grassland-scrubland. *Journal of Fish and Wildlife Management* 1, 152–160.

Bentley, W. W. (1959). The range relationships of Roosevelt elk, *Cervus canadensis roosevelti* (Merriam), at Prairie Creek Redwoods State Park, Humboldt County, California, in 1958 (Unpublished master's thesis). Humboldt State University, Arcata, CA.

Berger-Tal, O., & S. Bar-David. (2015). Recursive movement patterns: review and synthesis across species. *Ecosphere* 6, e149.

Bhaskar, G., & F. W. Weckerly. (2016). Birth mass scaling in elk (*Cervus elaphus*). *American Midland Naturalist* 175, 275–279.

Blackwell, B. F., T. W. Seamans, & T. L. DeVault. (2014). White-tailed deer response to vehicle approach: evidence of unclear and present danger. *Plos One* 9, e109988.

Bleich, V. C., J. D. Wehausen, & S. A. Holl. (1990). Desert-dwelling mountain sheep: conservation implications of a naturally fragmented distribution. *Conservation Biology* 4, 383–390.

Bliss, L. M., & F. W. Weckerly. (2016). Habitat use by male and female Roosevelt elk in northwestern California. *California Fish and Game* 102, 8–16.

Blood, D. A., & G. W. Smith. (1984). Weights and measurements of Roosevelt elk on Vancouver Island. *The Murrelet* 65, 41–44.

Bonenfant, C., J. M. Gaillard, S. Dray, A. Loison, M. Royer, & D. Chessel. (2007). Testing sexual segregation and aggregation: old ways are best. *Ecology* 88, 3202–3208.

Bowler, D. E., & T. G. Benton. (2005). Causes and consequences of animal dispersal strategies: relating individual behaviour to spatial dynamics. *Biological Review* 80, 205–225.

Bowyer, R. T. (1981). Activity, movement, and distribution of Roosevelt elk during rut. *Journal of Mammalogy* 62, 574–582.

———. (2004). Sexual segregation in ruminants: definitions, hypotheses, and implications for conservation and management. *Journal of Mammalogy* 85, 1039–1052.

Bowyer, R. T., J. G. Kie, & V. Van Ballengerghe. (1998). Habitat selection by neonatal black-tailed deer: climate, forage, or risk of predation? *Journal of Mammalogy* 79, 415–425.

Bowyer, R. T., V. C. Bleich, K. M. Stewart, J. C. Whiting, & K. L. Monteith. (2014). Density dependence in ungulates: a review of causes, and concepts with some clarifications. *California Fish and Game* 100, 550–572.

Box, G. E. P., W. G. Hunter, & J. S. Hunter. (1978). *Statistics for experimenters.* New York: John Wiley & Sons.

Boyce, M. S. (1989). *The Jackson elk herd: intensive wildlife management in North America.* Cambridge University Press, Cambridge University Press.

———. (1991). Migratory behavior and management of elk (*Cervus elaphus*). *Applied Animal Behaviour Science* 29, 239–250.

Brook, B. W., & C. J. A. Bradshaw. (2006). Strength of evidence for density dependence in abundance time series of 1198 species. *Ecology* 87, 1445–1451.

Brooks, S. P., & A. Gelman. (1998). General methods for monitoring convergence of iterative simulations. *Journal of Computational and Graphical Statistics* 9, 515–529.

Burnham, K. P., & D. R. Anderson. (2002). *Model selection and multimodel inference: a practical information-theoretic approach.* New York: Springer Verlag.

Byers, J. A. (1997). *American pronghorn: social organization and ghosts of predators past.* Chicago: University of Chicago Press.

California Department of Fish and Wildlife. n.d. Elk hunting. https://www.wildlife.ca.gov/Hunting/Elk#19519270-harvest-statistics

Caughley, G. (1970). Eruption of ungulate populations, with emphasis on Himalayan Thar in New Zealand. *Ecology* 51, 53–72.

———. (1976). Wildlife management and the dynamics of ungulate populations. In T. H. Coaker (Ed.), *Applied biology* (vol. 1, 183–246). New York: Academic Press.

———. (1994). Directions in conservation biology. *Journal of Animal Ecology* 63, 215–244.

Caughley, G., & A. R. E. Sinclair. (1994). *Wildlife ecology and management.* Cambridge, MA: Blackwell.

Church, J. S., & R. J. Hudson. (1996). Calving behaviour of farmed wapiti (*Cervus elaphus*). *Applied Animal Behaviour Science* 46, 263–270.

Clark, F., B. W. Brook, S. Delean, H. R. Akçakaya, & C. J. A. Bradshaw. (2010). The theta-logistic is unreliable for modelling most census data. *Methods in Ecology and Evolution* 1, 253–262.

Clutton-Brock, T. H., F. E. Guinness, & S. D. Albon. (1982). *Red deer: behavior and ecology of two sexes.* Chicago: University of Chicago Press.

Clutton-Brock, T. H., K. E. Rose, & F. E. Guinness. (1997). Density-related changes in sexual selection. *Proceedings of the Royal Society*, part B 264, 1509–1516.

Cobb, M. A. (2010). Spatial ecology and population dynamics of tule elk (*Cervus elaphus nannodes*) at Point Reyes National Seashore, California (Unpublished doctoral dissertation). University of California, Berkeley.

Cole, E. K., M. D. Pope, & R. G. Anthony. (2004). Influence of road management on diurnal habitat use of Roosevelt elk. *Northwest Science* 78, 313–321.

Cole, E. K., R. G. Anthony, & M. D. Pope. (1997). Effects of road management on movement and survival of Roosevelt elk. *Journal of Wildlife Management* 61, 1115–1126.

Collins, W. B., P. J. Urness, & D. D. Austin. (1978). Elk diets and activities on different lodgepole pine habitat segments. *Journal of Wildlife Management* 42, 779–810.

Coltman, D. W., J. G. Pilkington, & J. M. Pemberton. (2003). Fine-scale genetic structure in a free-living ungulate population. *Molecular Ecology* 12, 733–742.

Conroy, K. A. (1987). The effects of short duration grazing on the production and nutritional quality of forage on two sites in western Humboldt County, California (Unpublished master's thesis). Humboldt State University, Arcata, CA.

Cook, J. G., R. C. Cook, R. W. Davis, & L. L. Irwin. (2016). Nutritional ecology of elk during summer and autumn in the Pacific Northwest. *Wildlife Monographs* 195, 1–81.

Cook, R. C., J. G. Cook, D. J. Vales, B. K. Johnson, S. M. McCorquodale, L. A. Shipley, ... L. Schmitz. (2013). Regional and seasonal patterns of nutritional condition and reproduction in elk. *Wildlife Monographs* 184, 1–44.

Coulson, T., F. Guinness, J. Pemberton, & T. Clutton-Brock. (2004). The demographic consequences of releasing a population of red deer from culling. *Ecology* 85, 411–422.

Crawley, M. J. (1983). *Herbivory: the dynamics of animal-plant interactions.* New York: Blackwell Press.

Dale, H. C. (1918). *The Ashley-Smith explorations and the discovery of a central route to the Pacific, 1822–1829*. Cleveland, OH: Arthur H. Clark.

Dasmann, R. F. (1964). *Big game of the redwood forest region*. Arcata, CA, National Park Service.

Daubenmire, R. F. (1968). *Plant communities: a textbook of plant synecology*. New York: Harper & Row.

den Boer, P. J. (1981). On the survival of populations in a variable and heterogeneous environment. *Oecologia* 50, 39–50.

Dennis, B., & R. F. Costantino. (1988). Analysis of steady-state populations with the gamma abundance model: application to *Tribolium*. *Ecology* 69, 1200–1213.

Dennis, B., J. M. Ponciano, S. R. Lele, M. L. Taper, & D. F. Staples. (2006). Estimating density dependence, process noise, and observation error. *Ecological Monographs* 76, 323–341.

Dennis, B., & M. L. Taber. (1994). Density dependence in time series observations of natural populations: estimation and testing. *Ecological Monographs* 64, 205–224.

Duarte, A. (2015). Population dynamics and habitat conservation for the golden-cheeked warbler (*Setophaga chrysoparia*) (Unpublished doctoral dissertation). Texas State University, San Marcos.

Duarte, A., F. W. Weckerly, M. Schaub, & J. S. Hatfield. (2016). Estimating golden-cheeked warbler immigration: implications for the spatial scale of conservation. *Animal Conservation* 19, 65–74.

Dywernchuk, L. W., & D. A. Boag. (1972). Ducks nesting in association with gulls—an ecological trap? *Canadian Journal of Zoology* 50, 559–563.

Early, D. A. (2010). Intraspecific black bear spatial patterns and interactions at a small spatio-temporal scale (Unpublished master's thesis). Humboldt State University, Arcata, CA.

Eberhardt, L. E., L. L. Eberhardt, B. L. Tiller, & L. L. Cadwell. (1996). Growth of an isolated elk population. *Journal of Wildlife Management* 60, 369–373.

Ehrlich, P. R., & G. C. Daily. (1993). Population extinction and saving biodiversity. *Ambio* 22, 64–68.

Elbroch, L. M., P. E. Lendrum, J. Newby, H. Quigley, & D. Craighead. (2013). Seasonal foraging ecology of non-migratory cougars in a system with migrating prey. *Plos One* 8, e83375.

Ellsworth, R. S. (1930). Hunting elk for the market in the forties. *California Fish and Game* 16, 367.

Fifield, V. L., A. J. Rossi, & E. E. Boydston. (2015). Documentation of mountain lions in Marin County, California, 2010–2013. *California Fish and Game* 101, 66–71.

Forsyth, D. M., & P. Caley. (2006). Testing the irruptive paradigm of large-herbivore dynamics. *Ecology* 87, 297–303.

Fortin, D., M. S. Boyce, E. H. Merrill, & J. M. Fryxell. (2004). Foraging costs of vigilance in large mammalian herbivores. *Oikos* 107, 172–180.

Fowler, G. W. (1981). Density dependence as related to life history strategy. *Ecology* 62, 602–610.

Franklin, W. L. (1968). Herd organization, territoriality, movements and home range in Roosevelt elk (Unpublished master's thesis). Humboldt State University, Arcata, CA.

Franklin, W. L., & J. W. Lieb. (1979). The social organization of a sedentary population of North American elk: A model for understanding other populations. In M. S. Boyce & L. D. Hayden-Wing (Eds.), *North American elk: Ecology, behavior and management* (185–195). Laramie, University of Wyoming.

Franklin, W. L., A. S. Mossman, & M. Dole. (1975). Social organization and home range of Roosevelt elk. *Journal of Mammalogy* 56, 102–118.

Fritschle, J. A. (2008). Reconstructing historic ecotones using the public land survey: the lost prairies of Redwood National Park. *Annals of the Association of American Geographers* 98, 24–39.

Fryxell, J. M., M. Hazell, L. Borger, B. D. Dalziel, D. T. Haydon, J. M. Morales, ... R. C. Rosatte. (2008). Multiple movement modes by large herbivores at multiple satiotemporal scales. *Proceedings of the National Academy of Sciences (USA)* 105, 19114–19119.

Fryxell, J. M., J. F. Wilmshurst, A. R. E. Sinclair, D. T. Haydon, R. D. Holt, & P. A. Abrams. (2005). Landscape scale, heterogeneity, and the viability of Serengeti grazers. *Ecology Letters* 8, 328–335.

Gaillard, J. M., & N. G. Yoccoz. (2003). Temporal variation in survival of mammals: a case of environmental canalization. *Ecology* 84, 3294–3306.

Gaillard, J. M., M. Festa-Bianchet, & N. G. Yoccoz. (1998). Population dynamics of large herbivores: variable recruitment with constant adult survival. *Trends in Ecology and Evolution* 13, 28–63.

Gedir, J. V., & R. J. Hudson. (2000). Seasonal intake determination in reproductive wapiti hinds (*Cervus elaphus canadensis*) using n-alkane markers. *Canadian Journal of Animal Science* 80, 137–144.

Geist, V. (1986). New evidence of high frequency of antler wounding in cervids. *Canadian Journal of Zoology* 64, 380–384.

Georgii, B. (1981). Activity patterns of female red deer (*Cervus elaphus* L.) in the Alps. *Oecologia* 49, 127–136.

Gilpin, M. E., & F. J. Ayala. (1973). Global models of growth and competition. *Proceedings of the National Academy of Sciences, USA* 70: 3590–3593.

Gogan, P. J. P., & R. H. Barrett. (1987). Roosevelt elk dietary quality in northern coastal California. *California Fish and Game* 80, 80–83.

Green, E. K., & S. M. Galatowitsch. (2002). Effects of *Phalaris arundinacea* and nitrate-N addition on the establishment of wetland plant communities. *Journal of Applied Ecology* 39, 134–144.

Green, R. A., & G. D. Bear. (1990). Seasonal cycles of daily activity patterns of Rocky Mountain elk. *Journal of Wildlife Management* 54, 272–279.

Grenier, J. J., R. T. Golightly, Jr., & T. D. Hofstra. (1990). Habitat use by Roosevelt elk in Redwood National Park, California. *Proceedings of the Western States and Provinces Elk Workshop 1990*, 8–13.

Grenier, J. J. (1991). Habitat use by Roosevelt elk in the Bald Hills of Redwood National Park (Unpublished master's thesis). Humboldt State University, Arcata, CA.

Griffin, K. A., M. Hebblewhite, H. S. Robinson, P. Zager, S. M. Barber-Meyer, D. Christianson, . . . P. J. White. (2011). Neonatal mortality of elk driven by climate, predator phenology and predator community. *Journal of Animal Ecology* 80, 1246–1257.

Gross, J. E., L. A. Shipley, N. T. Hobbs, D. E. Spalinger, & B. A. Wunder. (1993). Functional-response of herbivores in food-concentrated patches: tests of a mechanistic model. *Ecology* 74, 778–791.

Gurevitch, J., & L. V. Hedges. (1993). In S. M. Scheiner and J. Gurevitch (Eds.). *Meta-analysis: combining the results of independent experiments. Design and analysis of ecological experiments* (378–398). New York: Chapman & Hall.

Guthrie, R. D. (1967). The extinct wapiti of Alaska and Yukon territory. *Canadian Journal of Zoology* 44, 47–57.

———. (1990). *Frozen fauna of the mammoth steppe: the story of Blue Babe.* Chicago: University of Chicago Press.

Hagman, J., G. Marten, & A. Hovin. (1975). Alkaloid concentration in plant parts of reed canarygrass of varying maturity. *Crop Science* 15, 41–43.

Hale, R., E. A. Treml, & S. E. Swearer. (2015). Evaluating the metapopulation consequences of ecological traps. *Proceedings of the Royal Society*, part B 282: 20142930.

Hammond, K. J., S. O. Hoskin, N. B. Jopson, C. G. Mackintosh, G. Hofstra, B. R. Thompson, & D. R. Stevens. (2013). Gastrointestinal tract development in red deer (*Cervus elaphus*) calves from 1 to 12 months of age. *Animal* 7, 1816–1823.

Harn, J. H. (1958). The Roosevelt elk, *Cervus canadensis roosevelti* (Merriam), at Prairie Creek Redwoods State Park, Humboldt County, California (Unpublished master's thesis). Humboldt State University, Arcata, CA.

Harper, J. A. (1961). Food habits and life history observations of the Roosevelt elk at Prairie Creek Redwoods State Park, Humboldt County, California (Unpublished master's thesis). Humboldt State University, Arcata, CA.

Harper, J. A., J. H. Harn, W. W. Bentley, & C. F. Yocom. (1967). The status and ecology of the Roosevelt elk in California. *Wildlife Monographs* 16, 1–49.

Harpole, J. L., & R. L. Lyman. (1999). The Holocene biogeographic history of elk (*Cervus elaphus*) in western Washington. *Northwest Science* 73, 106–112.

Hatfield, J. S., M. H. Reynolds, N. E. Seavy, & C. M. Krause. (2012). Population dynamics of Hawaiian seabird colonies vlunerable to sea-level rise. *Conservation Biology* 26, 667–678.

Heard, D. C. (1992). The effect of wolf predation and snow cover on musk-ox group size. *American Naturalist* 139, 190–204.

Hebblewhite, M., E. Merrill, & G. McDermid. (2008). A multi-scale test of the forage maturation hypothesis in a partially migratory ungulate population. *Ecological Monographs* 78, 141–166.

Hektner, M.M., R.W. Martin, & D.R. Davenport. (1983). The Bald Hills prairies of Redwood National Park. In C.I. Van Riper, L.D. Whittig, & M.L. Murphy (Eds.), *Proceedings of the first biennial conference of research in California's National Parks* (70–78). Davis, CA: University of California.

Hestbeck, J.B. (1982). Population regulation of cyclic mammals: the social fence hypothesis. *Oikos* 39, 157–163.

Hines, W.W., & J.C. Lemos. (1979). Reproductive performance by two age-classes of male Roosevelt elk in southwestern Oregon. Oregon Department of Fish and Wildlife, Portland.

Hirotani, A. (1990). Social organization of reindeer (*Rangifer tarandus*), with special reference to relationships among females. *Canadian Journal of Zoology* 68, 743–749.

Hofstra, T. (1997). Briefing information on Roosevelt elk in Redwood National and State parks (RNSP), prepared for Yurok Tribal Council. Arcata, CA, Redwood National Park.

Holling, C.S. (1973). Resilience and stability of ecological systems. *Annual Review of Ecology and Systematics* 4, 1–23.

Howell, J.A., G.C. Brooks, M. Semenoff-Irving, & C. Greene. (2002). Population dynamics of tule elk at Point Reyes National Seashore, California. *Journal of Wildlife Management* 66, 478–490.

Hudson, R.J., & M.T. Nietfeld. (1985). Effect of forage depletion on the feeding rate of wapiti. *Journal of Range Management* 38, 80–82.

Hudson, R.J., & W.G. Watkins. (1986). Foraging rates of wapiti on green and cured pastures. *Canadian Journal of Zoology* 64, 1705–1708.

Hunt, J.W., F.W. Weckerly, & J.R. Ott. (2012). Reliability of occupancy and N-mixture models for estimating abundance of golden-cheeked warblers (*Setophagia chrysoparia*). *Auk* 129, 105–114.

Husseman, J.S., D.L. Murray, G. Power, C. Mack, C.R. Wenger, & H. Quigley. (2013). Assessing differential prey selection patterns between two sympatric large carnivores. *Oikos* 101, 591–601.

Hutchings, J.A. (2015). Thresholds for impaired species recovery. *Proceedings of the Royal Society*, part B 282: 20150654.

Jarman, P.J. (1974). The social organisation of antelope in relation to their ecology. *Behaviour* 48, 215–267.

Jenkins, K.J., & E.E. Starkey. (1984). Social organization of Roosevelt elk in an old-growth forest. *Journal of Mammalogy* 63, 331–334.

———. (1991). Food habits of Roosevelt elk. *Rangelands* 13, 261–265.

Johnson, B.E., & J.H. Cushman. (2007). Influence of a large herbivore reintroduction on plant invasions and community composition in a California grassland. *Conservation Biology* 515–526.

Johnson, D.H. (2008). In defense of indices: the case of bird surveys. *Journal of Wildlife Management* 72, 857–868.

Julian, A. E., K. Schmidt, & F. W. Weckerly. (2013). Group size dynamics of female Roosevelt elk in Redwood National and State Parks. *California Fish and Game* 99, 49–54.

Kamler, J. F., B. Jedrzejewska, & W. Jedrzejewski. (2007). Activity patterns of red deer in Bialowieza National Park, Poland. *Journal of Mammalogy* 88, 508–514.

Kendall, B. E. (1998). Estimating the magnitude of environmental stochasticity in survival data. *Ecological Applications* 8, 184–193.

Kie, J. G., & R. T. Bowyer. (1999). Sexual segregation in white-tailed deer: density-dependent changes in use of space, habitat selection, and dietary niche. *Journal of Mammalogy* 80, 1004–1020.

Kirchhoff, M. D., & D. N. Larsen. (1998). Dietary overlap between native Sitka black-tailed deer and introduced elk in southeast Alaska. *Journal of Wildlife Management* 62, 236–242.

Klein, D. R. (1968). The introduction, increase, and crash of reindeer on St. Matthew Island. *Journal of Wildlife Management* 32, 350–367.

———. (1987). Vegetation recovery patterns following overgrazing by reindeer on St. Matthew Island. *Journal of Range Management* 40, 336–338.

Kolbe, N. R. (2016). Density-dependent not -independent factors influence Roosevelt elk recruitment in the Bald Hills of Redwood National Park (Unpublished master's thesis). San Marcos, Texas, Texas State University.

Kolbe, N. R., & F. W. Weckerly. (2015). Home-range overlap of Roosevelt elk herds in the Bald Hills of Redwood National Park. *California Fish and Game* 101, 208–217.

Komers, P. E., & G. P. Curman. (2000). The effect of demographic characteristics on the success of ungulate re-introductions. *Biological Conservation* 2000, 187–193.

Kristan, D. M. (1992). Behavioral ontogeny in Roosevelt elk calves (*Cervus elpahus roosevelti*) (Unpublished master's thesis). Humboldt State University, Arcata, CA.

Lande, R., S. Engen, & B. E. Saether. (2003). *Stochastic population dynamics in ecology and conservation*. Oxford, UK: Oxford University Press.

Le Saout, S., S. Chollet, S. Chamaille-Jammes, L. Blanc, S. Padie, T. Verchere, . . . J.-L. Martin. (2014). Understanding the paradox of deer persisting at high abundance in heavily browsed habitats. *Wildlife Biology* 20: 122–135.

Lee, D. E. (2015). Demography of giraffe in the fragmented Tarangire ecosystem (Unpublished doctoral dissertation). Dartmouth College, Hanover, NH.

Lemos, J. C. (1971). Roosevelt elk utilization of Redwood Forest cutovers in relation to regrowth and forest management (Unpublished master's thesis). Humboldt State University, Arcata, CA.

Lendrum, P. E., C. R. Anderson, K. L. Monteith, J. A. Jenks, & R. T. Bowyer. (2014). Relating the movement of a rapidly migrating ungulate to spatiotemporal patterns of forage quality. *Mammalian Biology* 79, 369–375.

Leopold, A. (1933). *Game management*. New York: C. Scribner's Sons.

Leopold, A. S. (1943). Deer irruptions. Transactions of the Wisconsin Academy of Sciences. *Arts and Letters* 35, 351–365.

Leopold, A. (1949). *A Sand County almanac: with essays on conservation.* New York: Oxford University Press.

Leopold, A. S. Jr., S. A. Cain, C. M. Cottam, I. N. Gabrielson, & T. L. Kimball. (1963). Wildlife management in the national parks. *Transactions of the North American Wildlife Conference* 24, 28–45.

Leslie, D. M. Jr. (1982). Nutritional ecology of cervids in old-growth forests in Olympic National Park, Washington (Unpublished doctoral dissertation). Corvallis, Oregon State University.

Levins, R. (1969). Some demographic and genetic consequences of environmental heterogeneity for biological control. *Bulletin of the Entomological Society of America* 15, 237–240.

Lieb, J. W. (1973). Social behavior in Roosevelt elk cow groups (Unpublished master's thesis). Humboldt State University, Arcata, CA.

Lindenmayer, D. B., & G. E. Likens. (2010). The science and application of ecological monitoring. *Biological Conservation* 143, 1317–1328.

Lindenmayer, D. B., G. E. Likens, C. J. Krebs, & R. J. Hobbs. (2010). Improved probability of detection of ecological "surprises." *Proceedings of the National Academy of Sciences (USA)* 107: 21957–21962.

Loe, L. E., A. Mysterud, V. Veiberg, & R. Langvatn. (2009). Negative density-dependent emigration of males in an increasing red deer population. *Proceedings of the Royal Society*, part B 2763: 2581–2587.

Logsdon, H. S. (1965). The immobilization and movements of the Roosevelt elk (Unpublished master's thesis). Humboldt State University, Arcata, CA.

Long, E. S., D. R. Diefenbach, C. S. Rosenberry, & B. D. Wallingford. (2010). Influence of roads, rivers, and mountains on natal dispersal of white-tailed deer. *Behavioral Ecology* 19, 1235–1242.

Lorenzini, R., & L. Garofalo. (2015). Insights into the evolutionary history of *Cervus* (Cervidae, tribe Cervini) based on Bayesian analysis of mitochondrial marker sequences, with first indications for a new species. *Journal of Zoological Systematics and Evolutionary Research* 53, 267–276.

Ludt, C. J., W. Schroeder, O. Rottmann, & R. Kuehn. (2004). Mitochondrial DNA phylogeography of red deer (*Cervus elaphus*). *Molecular Phylogenetics and Evolution* 31, 1064–1083.

MacArthur, R. H., & E. O. Wilson. (2001). *The theory of island biogeography.* New York: Oxford University Press.

Maier, J. A. K. (2001). Sex-specific dynamics of North American elk in relation to global climate. *Alces* 37, 411–420.

Mandel, R. D., & D. W. Kitchen. (1979). *The ecology of Roosevelt elk in and around Redwood National Park.* Arcata, CA: Redwood National Park.

Marshal, J. P., J. W. Cain, V. C. Bleich, & S. S. Rosenstock. (2009). Intrinsic and extrinsic sources of variation in the dynamics of large herbivore populations. *Canadian Journal of Zoology* 87, 103–111.

Marten, G., R. Barnes, A. Simons, & F. Wooding. (1973). Alkaloids and palatablility of *Phalaris arundinacea* L. grown in diverse environments. *Agronomy Journal* 65, 199–201.

Martina, J., & C. Von Ende. (2013). Increased spatial dominance in high nitrogen, saturated soil due to clonal architecture plasticity of the invasive wetland plant, *Phalaris arundinacea*. *Plant Ecology* 21, 1443–1453.

McCorquodale, S.M., L.L. Eberhardt, & L.E. Eberhardt. (1988). Dynamics of a colonizing elk population. *Journal of Wildlife Management* 52, 309–313.

McCullough, D.R. (1969). The tule elk: its history, behavior, and ecology. *University of California Publications in Zoology* 88, 1–191.

———. (1979). *The George Reserve deer herd: population ecology of a K selected species.* Ann Arbor: University of Michigan.

———. (1997). Irruptive behavior in ungulates. The science of overabundance: deer ecology and population management. W.J. McShea, H.B. Underwood, & J.H. Rappole. Washington, DC, Smithsonian Institution: 69–98.

———. (1999). Density dependence and life-history strategies of ungulates. *Journal of Mammalogy* 80, 1130–1146.

McLoughlin, P.D., M.S. Boyce, T. Coulson, & T. Clutton-Brock. (2006). Lifetime reproductive success and density-dependent, multi-variable resource selection. *Proceedings of the Royal Society*, part B 273: 1449–1454.

Meiri, M., A.M. Lister, M.J. Collins, N. Tuross, T. Goebel, S. Blockley,... I. Barnes. (2014). Faunal record identifies Bering isthmus conditions as constraint to end-Pleistocene migration to the new world. *Proceedings of the Royal Society*, part B 281: 2013–2167.

Meisingset, E., L.E. Loe, O. Brekkum, & A. Mysterud. (2014). Targeting mitigation efforts: the role of speed limit and road edge clearance for deer-vehicle collisions. *Journal of Wildlife Management* 78, 679–688.

Meredith, E.P., J.A. Rodzen, J.D. Banks, R. Schaefer, H.B. Ernest, T.R. Famula, & B.P. May. (2007). Microsatellite analysis of three subspecies of elk (*Cervus elaphus*) in California. *Journal of Mammalogy* 88, 801–808.

Merkle, J.A., D. Fortin, & J.M. Morales. (2014). A memory-based foraging tactic reveals an adaptive mechanism for restricted space use. *Ecology Letters* 17, 924–931.

Muñoz, J.D., A.O. Finley, R. Gehl, & S. Kravchenko. (2010). Nonlinear heirarchical models for predicting cover crop biomass using Normalized Difference Vegetation Index. *Remote Sensing of the Environment* 114, 2833–2840.

Murphy, A. (1963). A captive elk herd in Missouri. *Journal of Wildlife Management* 27, 411–414.

Mysterud, A., L.E. Loe, B. Zimmermann, R. Bischof, V. Veiberg, & E. Meisingset. (2011). Partial migration in expanding red deer populations at northern latitudes: A role for density dependence? *Oikos* 120, 1827–1825.

Nakagawa, S., & H. Schielzeth. (2013). A general and simple method for obtaining R^2 from genealized linear mixed-effects models. *Methods in Ecology and Evolution* 4, 133–142.

National Park Service. n.d. Your safety. http://www.nps.gov/redw/planyourvisit/safety.htm

Nichols, J. D., F. A. Johnson, B. K. Williams, & G. S. Boomer. (2015). On formally integrating science and policy: walking the walk. *Journal of Applied Ecology* 52, 539–543.

Noyes, J. H., B. K. Johnson, L. D. Bryant, F. S. Findholt, & J. W. Thomas. (1996). Effects of bull age on conception dates and pregnancy rates of cow elk. *Journal of Wildlife Management* 60, 508–517.

Orr, R. T. (1937). Notes on the life history of the Roosevelt elk in California. *Journal of Mammalogy* 18, 62–66.

Owen-Smith, N. (2010). *Dynamics of large herbivore populations in changing environments: towards appropriate models.* Hoboken, New Jersey: Wiley-Blackwell.

Peterson, L. P. (2016). Social factors during foraging bouts influence sexual segregation in Roosevelt elk (Unpublished master's thesis). San Marcos, Texas State University.

Pinheiro, J. C., & D. M. Bates. (2000). *Mixed-effects models in S and S-PLUS.* New York: Springer.

Polansky, L., P. de Valpine, J. O. Lloyd-Smith, & W. M. Getz. (2009). Likelihood ridges and multimodality in population growth rate models. *Ecology* 90, 2313–2320.

Pollard, J. C., G. W. Asher, & R. P. Littlejohn. (2002). Weaning date affects calf growth rates and hind conception dates in farmed red deer (*Cervus elaphus*). *Animal Science* 74, 111–116.

Pope, M. D. (1994). Roosevelt elk habitat use in the Oregon Coast Range (Unpublished master's thesis). Corvallis, Oregon State University.

R Core Team. (2014). *A language and environment for statistical computing.* Vienna, Austria: Author.

Raithel, J. D., M. J. Kauffman, & D. H. Pletscher. (2007). Impact of spatial and temporal variation in calf survival on the growth of elk populations. *Journal of Wildlife Management* 71, 795–803.

Ricca, M. A. (2013). Introduced caribou and reindeer in the Aleutian Archipelago of Alaska: irruptive population dynamics and interactions with aboveground-belowground ecosystem processess (Unpublished doctoral dissertation). University of California, Davis.

Ricca, M. A., D. H. Van Vuren, F. W. Weckerly, J. C. Williams, & A. K. Miles. (2014). Irruptive dynamics of introduced caribou on Adak Island, Alaska: an evaluation of Riney-Caughley model predictions. *Ecosphere* 5, a94.

Richey, R., & J. Wheeler. (2001). *Elk meadow: a panorama of history.* Orick, CA: Redwood National Park.

Riotte-Lambert, L., S. Benhamou, & S. Chamaille-Jammes. (2015). How memory-based movement leads to nonterritorial spatial segregation. *American Naturalist* 185, E103–E116.

Rutberg, A. T. (1986). Dominance and its fitness consequences in American bison cows. *Behaviour* 96, 62–91.

Saether, B. E. (1997). Environmental stochasticity and population dynamics of large herbivores: a search for mechanisms. *Trends in Ecology and Evolution* 12, 143–149.

Saether, B. E., M. Lillegard, V. Grotan, M. C. Drever, S. Engen, T. D. Nudds, & K. M. Podruzny. (2008). Geographical gradients in the population dynamics of North American prairie ducks. *Journal of Animal Ecology* 77, 869–882.

Sargeant, G. A., & M. W. Oehler Sr. (2007). Dynamics of newly established elk populations. *Journal of Wildlife Management* 71, 1141–1148.

Sauer, J. R., J. E. Fallon, & R. Johnson. (2003). Use of North American breeding bird survey data to estimate population change for bird conservation regions. *Journal of Wildlife Management* 67, 372–389.

Schindler, D. E., J. B. Armstrong, & T. E. Reed. (2015). The portfolio concept in ecology and evolution. *Frontiers in Ecology and the Environment* 13, 257–263.

Schmidt, K. (2015). *2014 herd unit classification and management of Roosevelt elk.* Arcata, CA: Redwood National Park.

Schmidt, R. H. (1991). Gray wolves in California: their presence and absence. *California Fish and Game* 77, 79–85.

Schwartz, J. E. II., & G. E. Mitchell. (1945). The Roosevelt elk of the Olympic Peninsula, Washington. *Journal of Wildlife Management* 9, 295–319.

Seavy, N. E., M. H. Reynolds, W. A. Link, & J. S. Hatfield. (2009). Postcatastrophe population dynamics and density dependence of an endemic island duck. *Journal of Wildlife Management* 73, 414–418.

Shelton, S. L., & F. W. Weckerly. (2007). Inconsistencies in historical geographic range maps: the gray wolf as example. *California Fish and Game* 93, 224–227.

Simard, M. A., S. D. Cote, R. B. Weladji, & L. J. Huot. (2008). Feedback effects of chronic browsing on life-history traits of a large herbivore. *Journal of Animal Ecology* 77, 678–686.

Sinclair, A. R. E., & J. P. Parkes. (2008). On being the right size: food-limited feedback on optimal body size. *Journal of Animal Ecology* 77, 635–637.

Sokal, R. R., & R. J. Rohlf. (2012). *Biometry: the principles and practice of statistics in biological research.* 4th ed. New York: W. H. Freeman and Co.

Sousa, W. P. (1984). The role of disturbance in natural communities. *Annual Review of Ecology and Systematics* 15, 353–391.

Staples, D. F., M. L. Taper, & B. B. Shepard. (2005). Risk-based viable population monitoring. *Conservation Biology* 19, 1908–1916.

Starns, H. D., M. A. Ricca, A. Duarte, & F. W. Weckerly. (2014). Climatic and density influences on recruitment in an irruptive population of Roosevelt elk. *Journal of Mammalogy* 95, 925–932.

Starns, H. D., F. W. Weckerly, M. A. Ricca, & A. Duarte. (2015). Vegetation changes associated with a population irruption by Roosevelt elk. *Ecology and Evolution* 5, 109–120.

Stevens, M. C. (1965). A preliminary study of the movement, distribution, abundance, and sex and age composition of the Roosevelt elk in and adjacent to the

Prairie Creek Redwoods State Park (Unpublished master's thesis). Humboldt State University, Arcata, CA.

Stewart, K. M., R. T. Bowyer, B. L. Dick, B. K. Johnson, & J. G. Kie. (2005). Density-dependent effects on physical condition and reproduction in North American elk: an experimental test. *Oecologia* 143, 85–93.

Stewart, K. M., D. R. Walsh, J. G. Kie, B. L. Dick, & R. T. Bowyer. (2015). Sexual segregation in North American elk: the role of density dependence. *Ecology and Evolution* 5, 709–721.

Storer, T. I., & L. P. Tevis, Jr. (1955). *California grizzly.* Berkeley: University of California Press.

Street, G. M., A. R. Rodgers, T. Avgar, & J. M. Fryxell. (2015). Characterizing demographic parameters across environmental gradients: a case study with Ontario moose (*Alces alces*). *Ecosphere* 6, e138.

Street, G. M., F. W. Weckerly, & S. Schwinning. (2013). Modeling forage mediated aggregation in a gregarious ruminant. *Oikos* 122, 929–937.

Sueur, C., A. J. King, L. Conradt, G. Kerth, D. Lusseau, C. Mettke-Hofmann, . . . F. Aureli. (2011). Collective decision-making and fission-fusion dynamics: a conceptual framework. *Oikos* 120, 1608–1617.

Sullivan, M. S. (1934). *The travels of Jedediah Smith.* Santa Ana, CA: Fine Arts Press.

Taber, M. L., & P. J. P. Gogan. (2002). The Northern Yellowstone elk: density dependence and climatic conditions. *Journal of Wildlife Management* 66, 106–122.

Taillon, J., P. S. Barboza, & S. D. Cote. (2013). Nitrogen allocation to offspring and milk production in a capital breeder. *Ecology* 94, 1815–1827.

Thouless, C. R., & F. E. Guinness. (1986). Conflict between red deer hinds: the winner always wins. *Animal Behaviour* 34, 1166–1171.

Tukey, Jr., H. B. (1970). The leaching of substances from plants. *Annual Review of Plant Physiology* 21, 305–324.

Tyler, N. J. C. (2010). Climate, snow, ice, crashes, and declines in populations of reindeer and caribou (*Rangifer tarandus* L.). *Ecological Monographs* 80, 197–219.

van Beest, F. M., E. Vander Wal, A. V. Stronen, P. C. Paquet, & R. K. Brook. (2013). Temporal variation in site fidelity: scale-dependent effects of forage abundance and predation risk in a non-migratory large herbivore. *Oecologia* 173, 409–420.

Van Horne, B. (1983). Density as a misleading indicator of habitat quality. *Journal of Wildlife Management* 47, 893–901.

Van Moorter, B., D. Visscher, S. Benhamou, & L. Borger. (2009). Memory keeps you at home: a mechanistic model for home range emergence. *Oikos* 118, 641–652.

Veirs, S. D. Jr. (1987). *Vegetation studies of elk prairie, Prairie Creek Redwoods State Park, Humboldt County, California.* Arcata, CA: Cooperative Park Studies Unit, National Park Service, Redwood National Park.

Wang, G., N. T. Hobbs, S. Twombly, R. B. Boone, A. W. Illius, I. J. Gordon, & J. E. Gross. (2009). Density dependence in northern ungulates: interactions with predation and resources. *Population Ecology* 51, 123–132.

Weaver, S. P., & F. W. Weckerly. (2011). Sex ratio estimates of Roosevelt elk using counts and Bowden's estimator. *California Fish and Game* 97, 130–137.

Weckerly, F. W. (1993). Intersexual resource partitioning in black-tailed deer: a test of the body-size hypothesis. *Journal of Wildlife Management* 57, 475–494.

———. (1996). Roosevelt elk along the Prairie Creek drainage: an evaluation of estimating abundance and herd composition. *California Fish and Game* 82, 175–181.

———. (1998). Sexual segregation and competition in Roosevelt elk. *Northwestern Naturalist* 79, 113–118.

———. (1999). Social bonding and aggression in female Roosevelt elk. *Canadian Journal of Zoology* 77, 1379–1384.

———. (2001). Are large male Roosevelt elk less social because of aggression? *Journal of Mammalogy* 82, 414–421.

———. (2005). Grass and supplemental patch selection by a population of Roosevelt elk. *Journal of Mammalogy* 86, 630–638.

———. (2007). Constant proportionality in the female segment of a Roosevelt elk population. *Journal of Mammalogy* 71, 773–777.

Weckerly, F. W., & J. A. Foster. (2010). Blind count surveys of white-tailed deer and population estimates using Bowden's estimator. *Journal of Wildlife Management* 74, 1367–1377.

Weckerly, F. W., K. McFarland, M. Ricca, & K. Meyer. (2004). Roosevelt elk density and social segregation: foraging behavior and females avoiding larger groups of males. *American Midland Naturalist* 152, 386–399.

Weckerly, F. W., M. A. Ricca, & K. P. Meyer. (2001). Sexual segregation in Roosevelt elk: cropping rates and aggression in mixed-sex groups. *Journal of Mammalogy* 82, 825–835.

Westcott, D. A., C. S. Fletcher, A. McKeown, & H. T. Murphy. (2012). Assessment of monitoring power for highly mobile vertebrates. *Ecological Applications* 22, 374–383.

White, G. C., & R. A. Garrott. (1990). *Analysis of wildlife radio-tracking data.* New York: Academic Press.

Wickstrom, M. L., C. T. Robbins, T. A. Hanley, D. E. Spalinger, & S. M. Parish. (1984). Food intake and foraging energetics of elk and mule deer. *Journal of Wildlife Management* 48, 1285–1301.

Williams, B. K., J. D. Nichols, & M. J. Conroy. (2002). *Analysis and management of animal populations: modeling, estimation, and decision making.* San Diego, CA: Academic Press.

Wilmshurst, J. F., J. M. Fryxell, & R. J. Hudson. (1995). Forage quality and patch choice by wapiti. *Behavioral Ecology* 6, 209–217.

Wintle, B. A., M. C. Runge, & S. A. Bekessy. (2010). Allocating monitoring efforts in the face of unknown unknowns. *Ecology Letters* 13, 1325–1337.

Wistar, I. J. (1937). *Autobiography of Isaac Jones Wistar, 1827–1905, half a century in war and peace.* Philadelphia, PA: Wistar Institute of Anatomy and Biology.

Wolcott, D. M., R. L. Reitz, & F. W. Weckerly. (2015). Biological and environmental influences on parturition date and birth mass of a seasonal breeder. *Plos One* 10, e0124431.

Wolf, M., J. Frair, E. Merrill, & P. Turchin. (2009). The attraction of the known: the importance of spatial familiarity in habitat selection in Wapiti *Cervus elaphus*. *Ecography* 32, 401–410.

Woods, D., & K. Clark. (1971). Genetic control and seasonal variation of some alkaloids in reed canarygrass. *Canadian Journal of Plant Science* 51, 323–329.

About the Author

Butch Weckerly is professor of biology in the Department of Biology and director of the Master of Science Wildlife Ecology program at Texas State University. He received his doctorate from the University of Memphis. After graduate school, Butch held various positions at the University of California, Berkeley, and at Humboldt State University. He has been a faculty member at Texas State University since 2002. His research in population ecology and wildlife management has ranged from California to Alaska along the west coast; to New Mexico, Texas, and Tennessee in the southwestern and southeastern United States. His research is primarily with vertebrate species, particularly ungulates. He has been studying ungulates for more than 30 years. His research into the behavior, ecology, and management of Roosevelt elk began more than 20 years ago, and is ongoing. Butch has also served on scientific committees advising on conservation issues for the Texas Parks and Wildlife Department, the U.S. Fish and Wildlife Service, and Edwards Aquifer Authority in San Antonio, Texas. He has taught undergraduate- and graduate-level courses in vertebrate zoology, wildlife ecology and management, and applied statistics for 28 years. Butch has also mentored a number of graduate students at the master's and doctoral levels.

Index

Page numbers in *italics* indicate illustrations. The letter "t" indicates table data.